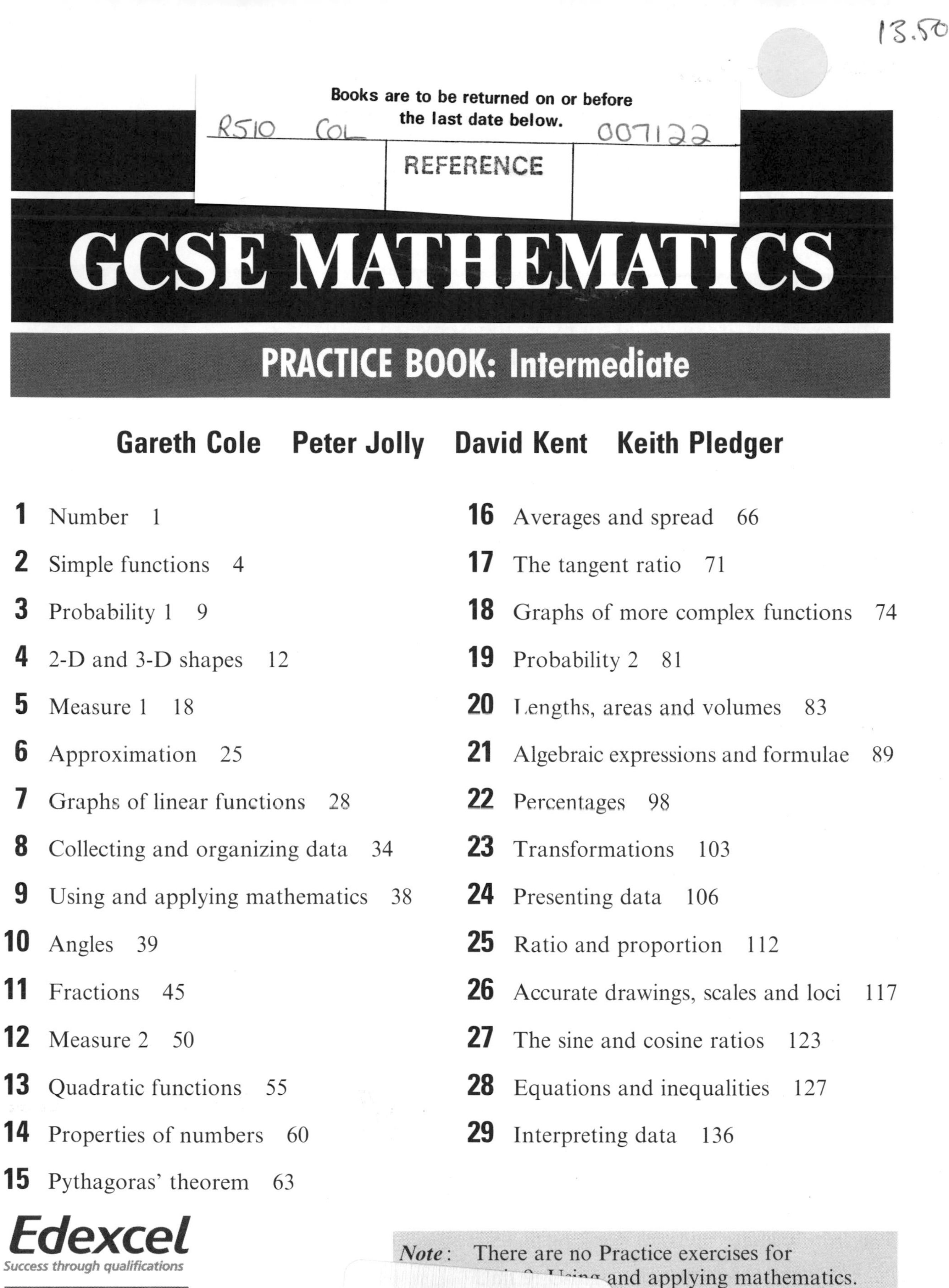

GCSE MATHEMATICS

PRACTICE BOOK: Intermediate

Gareth Cole   Peter Jolly   David Kent   Keith Pledger

Edexcel
Success through qualifications

Heinemann

Note: There are no Practice exercises for [illegible] and applying mathematics.

About this book

This book provides a substantial bank of additional exercises to complement those in the Edexcel GCSE Mathematics course textbook and offers a firm foundation for a programme of consolidation and homework.

Extra exercises are included for every topic covered in the course textbook, with the exception of the Using and applying and the Calculators and computers units.

Clear links to the course textbook exercises help you plan your use of the book:

Exercise 1.1 Links: (*1A, B*) 1A, B

This exercise is linked to exercises 1A and 1B in the old edition of the course textbook.

This exercise is linked to exercises 1A and 1B in the new edition of the course textbook.

Which edition am I using?

The new editions of the *Edexcel GCSE Maths* core textbooks have yellow cover flashes saying "ideal for the 2001 specification". You can also use the old edition (no yellow cover flash) to help you prepare for your exam.

Please note that the answers to the questions are provided in a separate booklet, available free when you order a pack of 10 practice books. You can buy further copies direct from Heinemann Customer Services.

Also available from Heinemann:

Edexcel GCSE Mathematics: Intermediate

The Intermediate textbook provides a complete course for the Intermediate tier examination.

Revise for Edexcel GCSE: Intermediate

The Intermediate revision book provides a structured approach to pre-exam revision and helps target areas for review.

1 Number

Exercise 1.1 Links: (*1A, B*) 1A, B

1 Write the numbers in words.

(a) 3271 **(b)** 8090 **(c)** 709
(d) 10 346 **(e)** 59 037 **(f)** 491 072
(g) 1 000 907 **(h)** 330 303 **(i)** 2 494 094

2 Write the numbers with digits.

(a) five hundred and forty nine
(b) fourteen thousand, seven hundred and seven
(c) three hundred and ninety five thousand and seven
(d) twelve million, one hundred and eight thousand, two hundred and three
(e) three and three quarter million
(f) two and a half thousand
(g) seven hundred and eighty six million, one hundred and forty five thousand, three hundred and fifty six

3 Make a place value table with these headings:
You will need room for twelve answers.
Put each of the decimals into the table and write them in their separate parts.

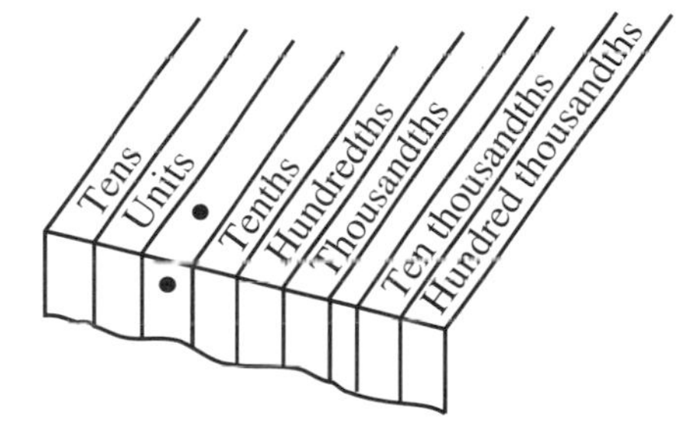

(a) 59.06 **(b)** 3.42
(c) 81.9 **(d)** 19.467
(e) 0.404 04 **(f)** 0.009 19
(g) 24.126 **(h)** 99.091 76
(i) 10.101 01 **(j)** 8.421 35
(k) 17.017 07 **(l)** 6.609 16

Exercise 1.2 Links: (*1C, D, E*) 1C, D, E

1 Write down the answers to these calculations.
Do **not** use a calculator.

(a) 23.3×100 **(b)** 9.19×10 **(c)** $46.7 \div 10$
(d) 0.127×100 **(e)** $0.017 \div 10$ **(f)** 7.56×1000
(g) $48.129 \div 1000$ **(h)** 79×100 **(i)** $36 \div 1000$
(j) $9.49 \div 100$ **(k)** 0.404×100 **(l)** $0.063 \div 100$
(m) 57.4×1000 **(n)** $0.0061 \div 10$ **(o)** $0.007\,14 \times 1000$

2 Go back to question **1** and use a calculator to check your answers.

3 Rearrange the measurements in each question in order of size, largest first.
(a) 7.01 m, 7.101 m, 7.0101 m, 7.4001 m
(b) 10.12 mm, 100.21 mm, 10.46 mm, 100.012 mm
(c) 9.7 kg, $10\frac{1}{2}$ kg, $9\frac{3}{4}$ kg, 9.9 kg
(d) 5.19 tonne, 9.51 tonne, 5.91 tonne, 5.59 tonne
(e) $6\frac{1}{2}$ pints, 6.6 pints, $5\frac{3}{4}$ pints, 6.2 pints, 6 and a quarter pints
(f) 78.09 m, 78.99 m, 78.96 m, 78.19 m, 78.009 m
(g) 0.167 *cl*, 0.161 *cl*, 0.919 *cl*, $\frac{3}{4}$ *cl*, 0.0167 *cl*

4 Work out these without using a calculator.

(a) 234×16	**(b)** 419×25	**(c)** 978×76
(d) $3088 \div 16$	**(e)** $2067 \div 53$	**(f)** $1258 \div 34$
(g) 719×59	**(h)** $5781 \div 47$	**(i)** $10\,001 \div 73$
(j) 89×843	**(k)** 563×97	**(l)** $16\,892 \div 82$

5 A van delivers 179 crates of cola. There are 24 bottles in a crate. How many bottles are there altogether?

6 A coach carries 56 passengers. How many coaches are needed to carry 1624 people?

Exercise 1.3 Links: (*1F–J*) 1F–J

For questions **1** to **3** draw a number line going from −10°C to +10°C to help you with these questions.

+10°
+9°
+8°
+7°
+6°
+5°
+4°
+3°
+2°
+1°
0°
−1°
−2°
−3°
−4°
−5°
−6°
−7°
−8°
−9°
−10°

1 Find the number of degrees between each pair of temperatures.

(a) 2°C, −3°C	**(b)** 9°C, −1°C	**(c)** −3°C, 5°C
(d) 8°C, −5°C	**(e)** −4°C, −10°C	**(f)** 0°C, −8°C
(g) 1°C, 8°C	**(h)** −1°C, 9°C	

2 Find the temperature when:

(a) 0°C rises by 4°C	**(b)** −3°C falls by 5°C
(c) 7°C falls by 12°C	**(d)** −2°C falls by −4°C
(e) 5°C falls by 9°C	**(f)** −1°C falls by −6°C

3 Rearrange the temperatures in each question in numerical order, lowest temperature first:
(a) −3°C, 5°C, 7°C, −1°C, 0°C, 3°C, −7°C, −4°C
(b) 9°C, −4°C, −9°C, 4°C, −3°C, 2°C, −2°C, 3°C
(c) −1°C, −10°C, −5 °C, 5°C, 3°C, 10°C, 8°C, −3°C

4 Work out these:

(a) $-5 - -5$	**(b)** $+7 + -3$	**(c)** $-9 - +6$	**(d)** $+2 + +2$
(e) $-7 + -3$	**(f)** $+14 - -8$	**(g)** $-1 + -4$	**(h)** $+5 - -7$
(i) $9 - 0$	**(j)** $+9 + -10$	**(k)** $-7 + +13$	**(l)** $+3 - -8$
(m) $-9 - -16$	**(n)** $-4 - +14$	**(o)** $+15 + +8$	**(p)** $-9 - +13$

5 Work out:

(a) $+2 \times +3$	**(b)** -2×-4	**(c)** $-3 \times +4$	**(d)** $+5 \times -3$
(e) $+15 \div -3$	**(f)** $+16 \div +4$	**(g)** $-20 \div -2$	**(h)** $+9 \div -1$
(i) -6×-7	**(j)** -8×-4	**(k)** $+6 \times -4$	**(l)** $+24 \div -3$
(m) $-30 \div -10$	**(n)** $+9 \times -7$	**(o)** $-48 \div +4$	**(p)** $-28 \div -2$
(q) $+9 \times -3$	**(r)** -7×-4	**(s)** $+96 \div -6$	**(t)** $-56 \div -8$

Exercise 1.4 Links: (*1K*) 1K

1 Write in words the number 90 909.

2 Write in digits the number eighty four thousand and ninety two.

3 Write these numbers in order of size, largest first.

69 696, 1 001 001, 596 695, 99 999, 490 904

4 Write these measurements in order of size, largest first:

4.051 mm, 4.501 mm, 4.5 mm, 4.05 mm, 4.005 mm

5 Fifty two people travel from Kingsley to London on a train. The train fare is £10.25 each. What is the total cost?

6 What is £1183.50 ÷ 45? Do **not** use a calculator.

7 Work out:

(a) $+5 - +1$	**(b)** $-9 - -7$	**(c)** $-12 \div -3$
(d) $-18 \div +6$	**(e)** -4×-4	**(f)** $-8 \div -8$

8 Work out 487 × 53 **without** using a calculator.

9 Work out 2744 ÷ 49 **without** using a calculator.

10 An airplane flying at 4500 feet climbs another 6900 feet and then descends by 8500 feet. At what height is the plane now?

2 Simple functions

Exercise 2.1 Links: (2*A* – *E*) 2A – E

1 The instructions for cooking a turkey are:

allow 45 minutes per kilogram then add an extra 30 minutes.

Work out the cooking time for a turkey that weighs:

(a) 3 kg **(b)** 8 kg **(c)** 12 kg
(d) 6.5 kg **(e)** 13.5 kg **(f)** 17.4 kg

2 **(a)** Draw a two-box number machine for these instructions:

take a number, multiply it by 5 then subtract 2.

(b) Put each number into your number machine and calculate what comes out:

(i) 3 **(ii)** 1 **(iii)** 8 **(iv)** 0 **(v)** 0.4
(vi) 2.6 **(vii)** 3.8 **(viii)** −2 **(ix)** −6 **(x)** 2.45

3 **(a)** Draw a two-box number machine for these instructions:

take a number, add 2 then multiply by 3.

(b) Put each number into your number machine and calculate what comes out:

(i) 1 **(ii)** 5 **(iii)** 0 **(iv)** 0.5 **(v)** 2.5
(vi) 1.35 **(vii)** −3 **(viii)** −2 **(ix)** −8 **(x)** −3.6

4 Here is a number machine.

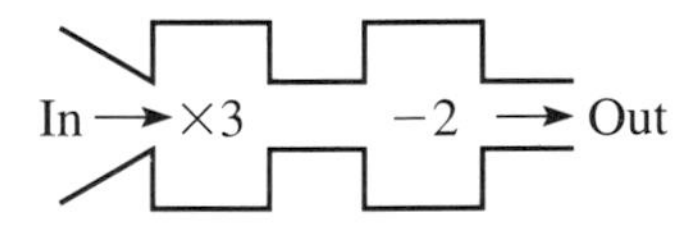

(a) Draw its inverse number machine.
(b) Work out the numbers that went into the original number machine for these numbers to come out:

(i) 10 **(ii)** 19 **(iii)** −2 **(iv)** 1 **(v)** 28
(vi) −20 **(vii)** 5.5 **(viii)** −5 **(ix)** −17 **(x)** 11.8

5 Here is a number machine.

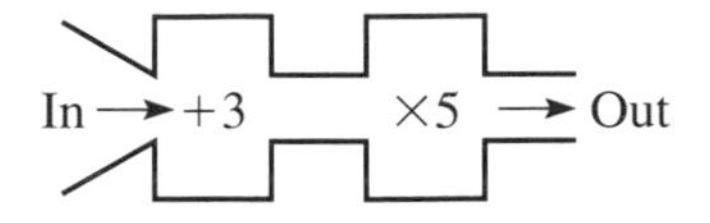

(a) Draw its inverse number machine.
(b) What numbers went into the machine for these numbers to come out?

(i) 25 **(ii)** 20 **(iii)** 35 **(iv)** 65 **(v)** 15
(vi) 0 **(vii)** 5 **(viii)** −20 **(ix)** 17.5 **(x)** −7.5

6 Put the numbers 0, 1, 2, 3, 4 and 5 into each of the six number machines.
Use the numbers that come out of the machines to help you find and write down number machines that are equivalent.

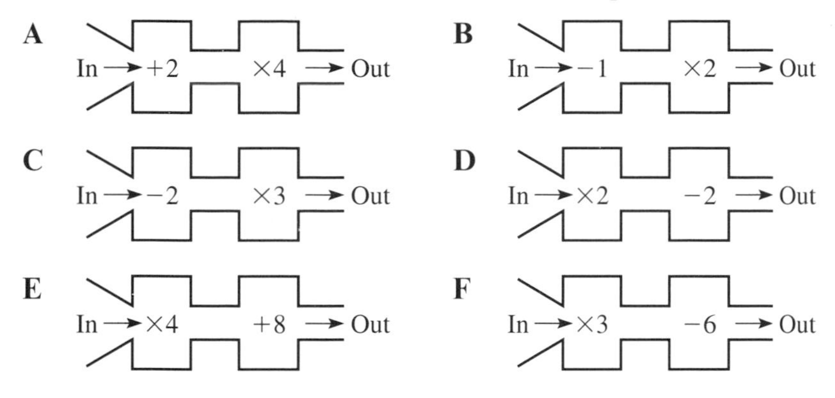

Exercise 2.2 Links: (2*F*–*H*) 2F–H

1 Write down the functions for each of these number machines:

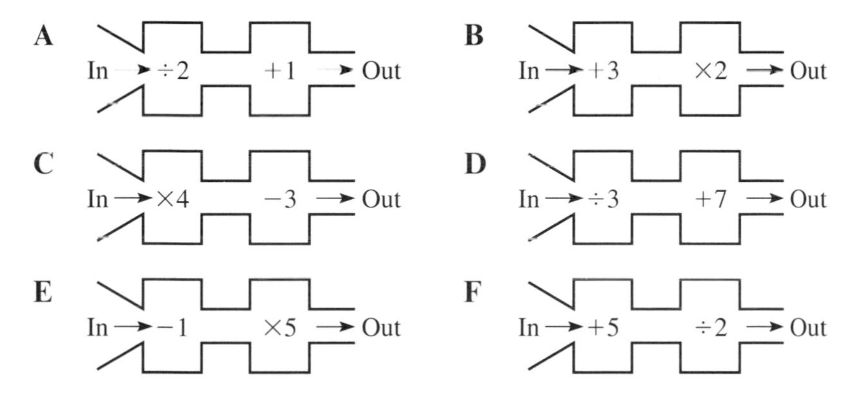

2 Draw the number machines for each of these functions:

(a) $n \rightarrow 2n + 3$ **(b)** $n \rightarrow 3n - 2$ **(c)** $n \rightarrow 5n + 1$

(d) $n \rightarrow -3n + 4$ **(e)** $n \rightarrow \frac{n}{3} + 4$ **(f)** $n \rightarrow 3(n + 2)$

(g) $n \rightarrow \frac{1}{2}(n + 5)$ **(h)** $n \rightarrow \frac{n + 3}{4}$ **(i)** $n \rightarrow 5(n - 3)$

3 Work out the outputs for each function using the inputs

$n = 1, 2, 3, 10$

(a) $n \rightarrow 3n - 2$ **(b)** $n \rightarrow n + 5$ **(c)** $n \rightarrow 2n + 7$

(d) $n \rightarrow 2(n + 5)$ **(e)** $n \rightarrow \frac{n}{2} + 5$ **(f)** $n \rightarrow 2(n - 1)$

(g) $n \rightarrow 2n - 2$ **(h)** $n \rightarrow \frac{n + 5}{2}$

4 Work out the outputs to each of the functions for the given inputs:

(a) $n \to n+3$ for $n = 0, 2, 5, 17$
(b) $n \to 3n+1$ for $n = 1, 3, 5, 7$
(c) $n \to 2n-3$ for $n = 0, 1, 2, 3, 4$
(d) $n \to 3n+2$ for $n = -2, -1, 0, 1, 2$
(e) $n \to 2(n+3)$ for $n = 0, 1, 2, 3$
(f) $n \to 2n+6$ for $n = 0, 1, 2, 3$

What do you notice about the outputs for the functions in **(e)** and **(f)**?

5 Here are six functions:

A $n \to n+2$ **B** $n \to 2n+4$ **C** $n \to 5(n-1)$
D $n \to 2(n+2)$ **E** $n \to 5n-5$ **F** $n \to 4n+2$

Write down the letters of the pairs of functions that are equivalent.

6 Write down a function equivalent to:

(a) $n \to 3(n+4)$ **(b)** $n \to 2n-10$ **(c)** $n \to \dfrac{3n+3}{3}$

Exercise 2.3 Links: (*2I–K*) 2I–K

1 Draw the inverse number machines for these functions and use them to work out the missing inputs and outcomes:

(a) $n \to 2n+1$
$3 \to$
$10 \to$
$\to 7$
$\to 11$

(b) $n \to 3n-2$
$2 \to$
$5 \to$
$\to 10$
$\to 16$

(c) $n \to 5n+3$
$5 \to$
$-1 \to$
$\to 13$
$\to 15$

(d) $n \to 4n-5$
$0 \to$
$4 \to$
$\to 23$
$\to -13$

(e) $n \to 4(n+1)$
$2 \to$
$0 \to$
$\to 20$
$\to -6$

(f) $n \to 2(n-5)$
$8 \to$
$2 \to$
$\to 2$
$\to -6$

(g) $n \to \dfrac{n}{2}+1$
$6 \to$
$10 \to$
$\to 5$
$\to 6$

(h) $n \to \dfrac{n+1}{5}$
$9 \to$
$16 \to$
$\to 5$
$\to -12$

2 For each of these functions, work out the input:

(a) $n \to n + 2$ when the output is 8

(b) $n \to 3n - 7$ when the output is 5

(c) $n \to \frac{n}{2} + 6$ when the output is 10

(d) $n \to 5(n - 2)$ when the output is 20

(e) $n \to \frac{n + 7}{3}$ when the output is 4

(f) $n \to 2n - 3$ when the output is 6

3 Write each of these functions in the form $n \to bn + c$

(a) $n \to 2(n + 1)$ **(b)** $n \to 3(n - 4)$

(c) $n \to 4(n - 2)$ **(d)** $n \to \frac{1}{2}(n + 6)$

(e) $n \to -(n + 5)$ **(f)** $n \to -2(n - 3)$

(g) $n \to \frac{2n + 8}{4}$ **(h)** $n \to \frac{3(n + 4)}{6}$

4 Here are three functions:

$n \to 2n - 6$ $n \to 2(n - 3)$ $n \to \frac{6n - 9}{3}$

Explain fully whether:

(a) they are all different

(b) two of them are equivalent (identify which two)

(c) they are all equivalent.

5 Find the function to fit each pattern of inputs and outputs:

(a)	(b)	(c)	(d)	(e)
$0 \to 0$	$0 \to 2$	$1 \to 4$	$6 \to 0$	$0 \to -5$
$1 \to 4$	$1 \to 5$	$3 \to 8$	$8 \to 1$	$4 \to 7$
$2 \to 8$	$2 \to 8$	$5 \to 12$	$10 \to 2$	$5 \to 10$
$3 \to 12$	$3 \to 11$	$7 \to 16$	$12 \to 3$	$8 \to 19$

Exercise 2.4 Links: (*2L*) 2L

1 Write down, in terms of n, an expression for the nth term in each of these sequences:

(a) 5, 8, 11, 14, 17 ...

(b) 3, 7, 11, 15, 19 ...

(c) 6, 12, 18, 24, 30 ...

(d) 4, 9, 14, 19, 24 ...

(e) 4, 1, −2, −5, −8 ...

(f) 37, 40, 43, 46, 48 ...

(g) 17, 23, 29, 35, 41 ...

2 The first five numbers in a sequence are:

4, 7, 10, 13, 16

(a) Find an expression in n for the nth term in this sequence.
(b) Use your expression to work out the 100th number in the sequence.
(c) Which term in the sequence is equal to 61?

3 Here is a pattern made out of matchsticks:

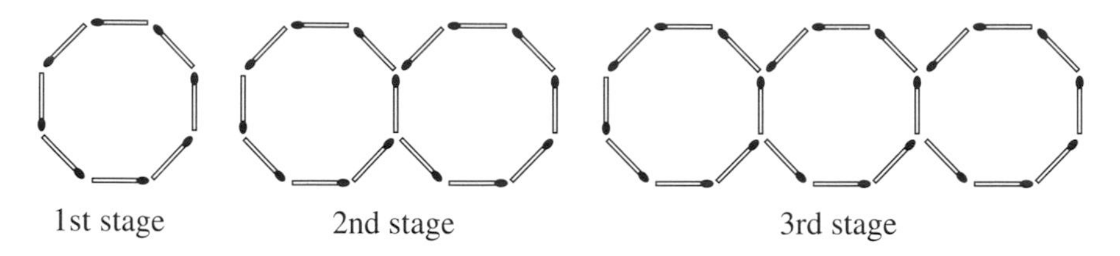

Find an expression for the number of matchsticks in the nth stage.

4 Rods can be fixed together using bolts. The diagrams show rods fixed together to form a pattern of hexagons in a row.

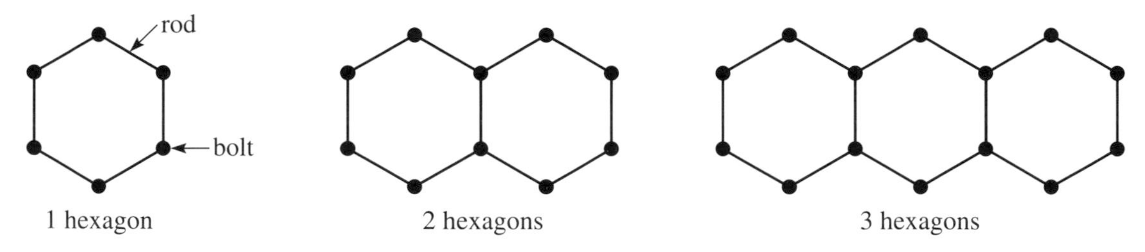

(a) Copy and complete the diagram below to show how rods can be fixed together to form a pattern of 4 hexagons in a row.

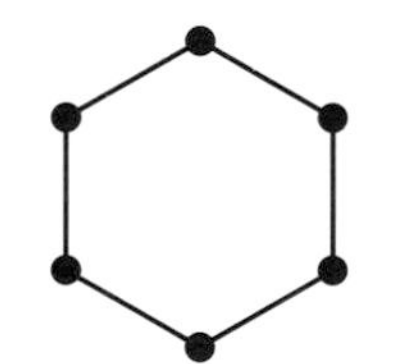

(b) Copy and complete the table.

Number of hexagons	1	2	3	4	5	6	7
Number of bolts	6	10	14				

(c) Work out the number of bolts needed to form a pattern of 15 hexagons in a row.
(d) Write down an expression, in terms of n, for the number of bolts needed to form a pattern of n hexagons in a row.

[E]

3 Probability 1

Exercise 3.1 Links: (*3A*) 3A

1 Draw a probability scale. Mark on it an estimate for the probability that:
(a) there is no sugar on sale at the supermarket
(b) New Year's Day will be on the 1st January
(c) the next set of traffic lights on a journey will be at red
(d) a packet of Cornflakes will contain Rice Krispies
(e) a wasp sting will result in death
(f) a Premiership football match will end as a draw.

2 On a probability scale mark the probabilities:
(a) 0.7 **(b)** 0.05 **(c)** $\frac{1}{4}$ **(d)** 35% **(e)** $\frac{4}{5}$

3 Events are classified as:

likely, very likely, quite likely, very unlikely,
unlikely, impossible, certain.

Mark these on a probability scale.

4 Put each list of probabilities in order of size starting with the least likely:
(a) $\frac{2}{5}$, 0.3, 20%
(b) 74%, 0.075, $\frac{3}{4}$
(c) 0.42, 40%, $\frac{4}{11}$

Exercise 3.2 Links: (*3B*) 3B

1 Two cubes each have 3 blue faces, 2 red faces and 1 green face.
List all the possible outcomes.
(Note: red, blue is the same as blue, red).
Which of these outcomes do you think is most likely? Which is the least likely?

Colour two dice and carry out an experiment to see if this confirms your answers.

2 Two dice each have 2 blue faces, 2 red faces, 1 green face and 1 yellow face.
List all the possible outcomes. Which outcome do you think is most likely? Which is least likely?

You can test your answer by an experiment.

3 Throw two ordinary dice. Each time record the score difference. Make a record of your results.

In a game, a difference of 1 or 2 is a win. Any other score is a loss. Is this fair? Use the results of your experiment to help you decide.

Exercise 3.3 Links: (*3C*) 3C

1 An eight-sided spinner has 2 red, 2 green, 3 blue and 1 yellow edge. What is the probability of obtaining:
(a) blue **(b)** yellow **(c)** not green **(d)** not blue?

2 Another eight-sided spinner has edges numbered 1 to 8. What is the probability of getting:
(a) an odd number **(b)** a square number
(c) a number greater than 5 **(d)** a number smaller than 3
(e) a prime number?

3 Using the spinner shown, what is the probability of
(a) winning
(b) not losing
(c) a draw?

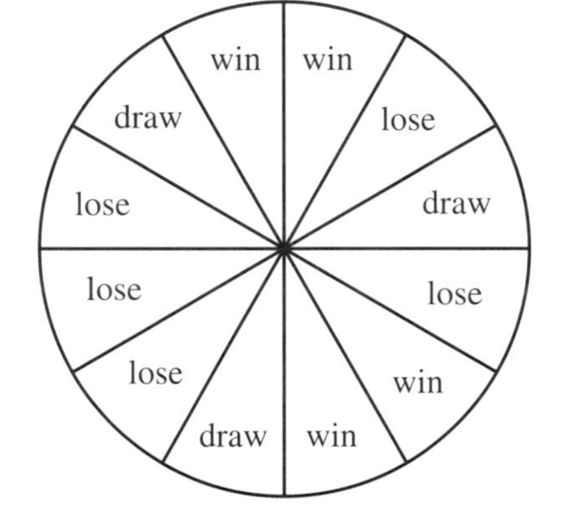

4 A letter is chosen at random from the word MISSISSIPPI. What is the probability that it is a:
(a) I **(b)** S **(c)** P **(d)** consonant **(e)** M?

5 A box of replacement lights for a Christmas set has 5 plain, 8 red, 3 blue and 4 green lights. If one is selected at random what is the probability that it will be:
(a) red **(b)** blue **(c)** plain **(d)** green **(e)** yellow

6 On a bookshelf there are 6 mathematics books, 3 history books, 2 geography books and 4 science books. If a book is taken at random what is the probability that it will:
(a) be a history book **(b)** not be a mathematics book
(c) be a science or a mathematics book?

7 Anne is asked to think of a number between 1 and 100 inclusive. Assuming her choice is random, what is the probability that her choice is:
(a) 10 or less **(b)** higher than 80 **(c)** an even number
(d) a square number **(e)** 23 **(f)** a cube number

Exercise 3.4 Links: (*3D*) 3D

1 A fair dice is rolled. What is the probability of getting:
(a) a multiple of 3 **(b)** a square number
(c) a number other than 4 **(d)** 9

2 A spinner has 12 equal sections. The probability of getting green is $\frac{1}{4}$; yellow is $\frac{1}{6}$; and blue is $\frac{5}{12}$. The other sections are all coloured red. How many red sections are there?

3 The nests of 40 birds are surveyed early in the breeding season. The results are:

Number of eggs	Blackbird	Thrush
1	2	4
2	5	9
3	9	8
4	2	0
5	1	0

One of these nests is visited at random.
(a) What is the probability that it is a blackbird's nest?
(b) What is the probability of the nest having exactly 3 eggs?
(c) If the nest visited is a blackbird's, what is the probability that it contains 5 eggs?

4 A T.V. can receive 5 channels.
When it is switched on it tunes in to the programme that was on previously.
The probability of it being switched off when tuned to the various channels are shown in the table.

BBC 1	BBC 2	ITV	Channel 4	Channel 5
	0.1	0.5	0.1	0.1

(i) What is the probability of getting BBC1 when the set is turned on?
(ii) What is the probability of getting an advertising channel?

4 2-D and 3-D shapes

Exercise 4.1 Links: (*4A*) 4A

1 Write down the mathematical names of these 2-D shapes:

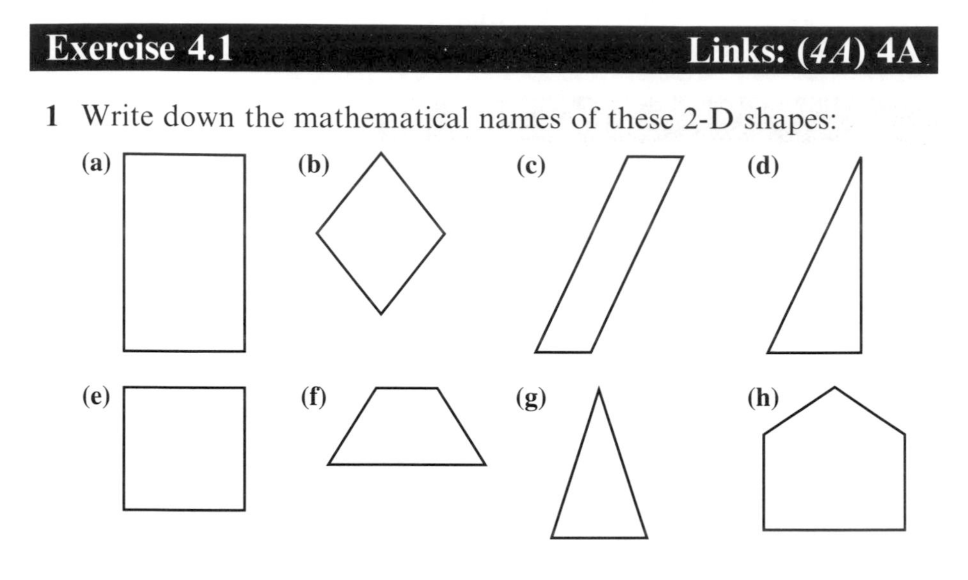

2 Write down the mathematical names of these 3-D shapes:

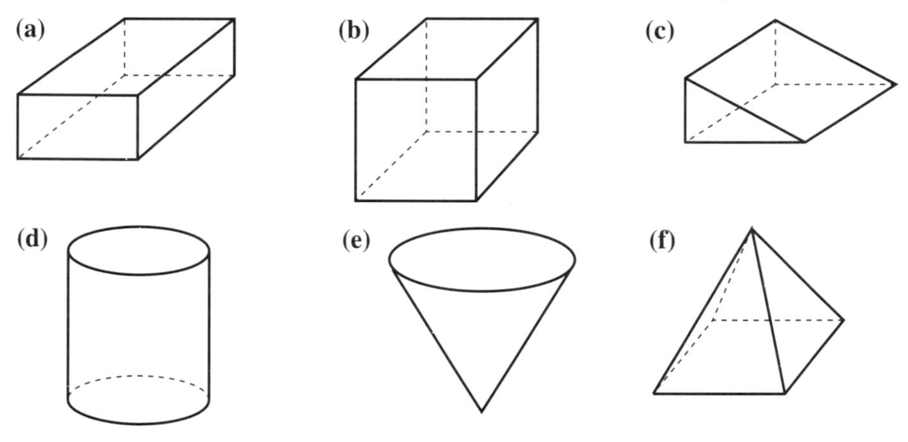

3 Write down the special names of these shapes:
- **(a)** a triangle with no sides equal
- **(b)** a triangle with all its angles equal
- **(c)** a triangle with two equal angles
- **(d)** a quadrilateral with one pair of parallel sides
- **(e)** a quadrilateral with two pairs of parallel sides and equal diagonals
- **(f)** a quadrilateral with two pairs of parallel sides and diagonals that are not equal

4 Write down the names of all the quadrilaterals that have:
- **(a)** two pairs of opposite angles equal
- **(b)** one pair of opposite angles equal
- **(c)** two pairs of adjacent angles equal

5 Calculate all the unlabelled angles in these shapes:

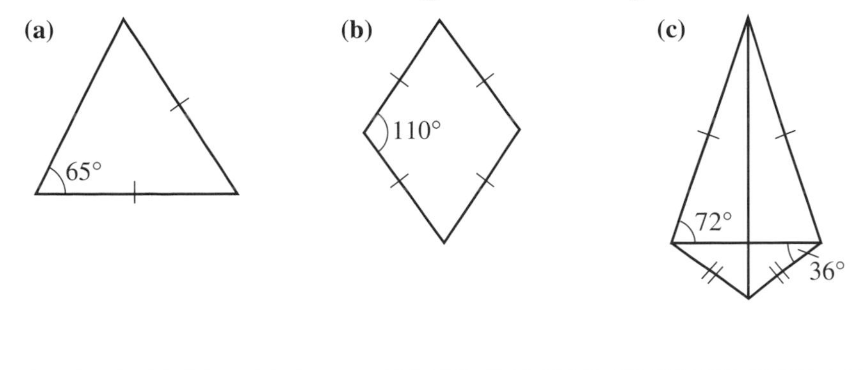

Exercise 4.2 Links: (*4B*) 4B

1 Write down the letters of the shapes in this diagram that are congruent.

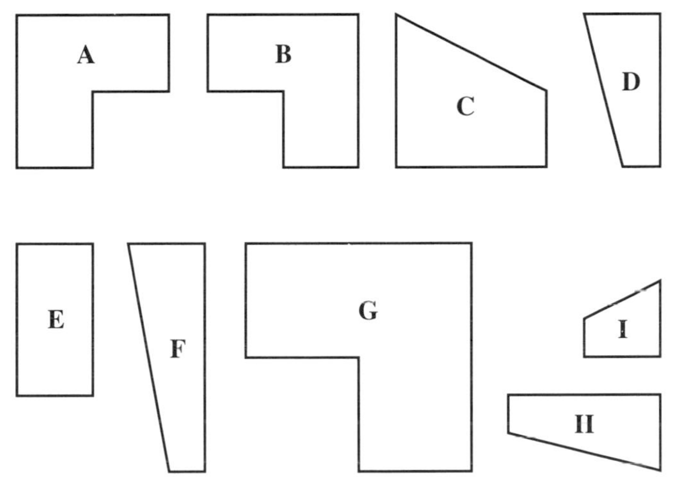

2 Write down the letters of the shapes in question **1** that are similar.

Exercise 4.3 Links: (*4C, D*) 4C, D

1 Draw, using tracing paper, each of these shapes. Draw on them any lines of symmetry that they might have.

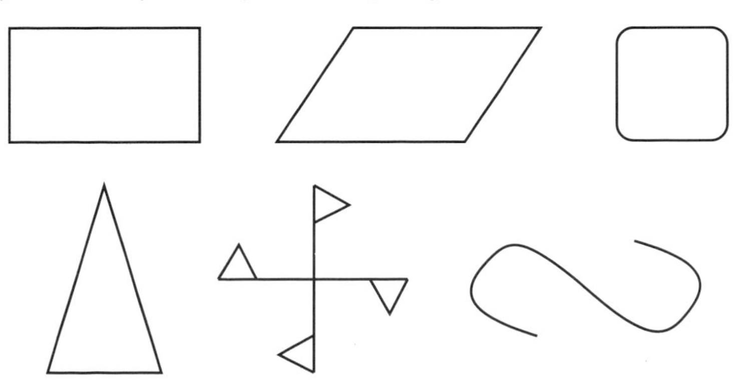

2 Write down the order of rotational symmetry for each of the shapes in question **1**.

3 Draw 2 copies of each of these 3-D shapes.
On each of them draw a plane of symmetry.

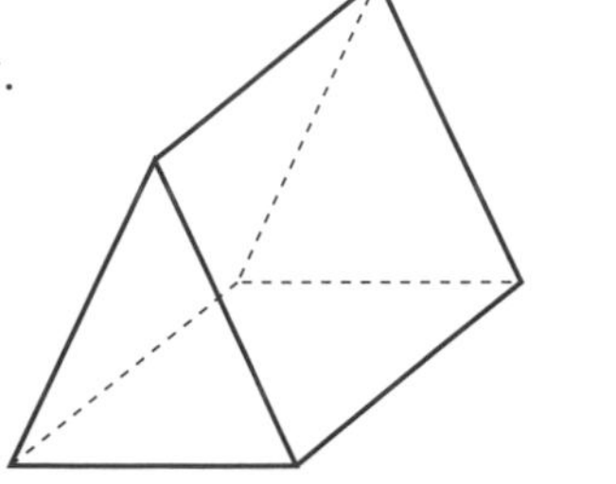

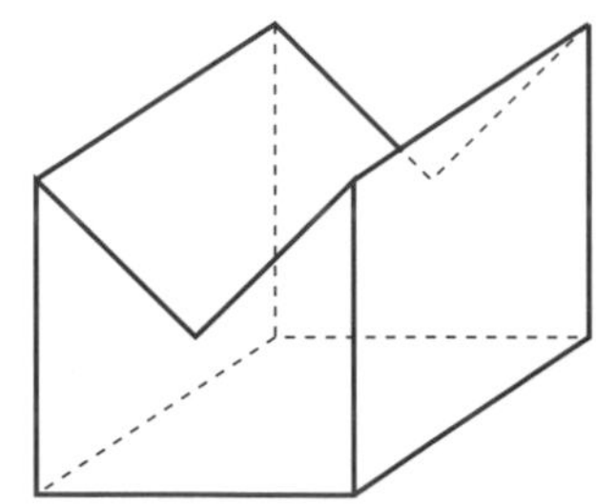

Exercise 4.4 Links: (*4E*) 4E

1 Draw a net of these 3-D shapes.

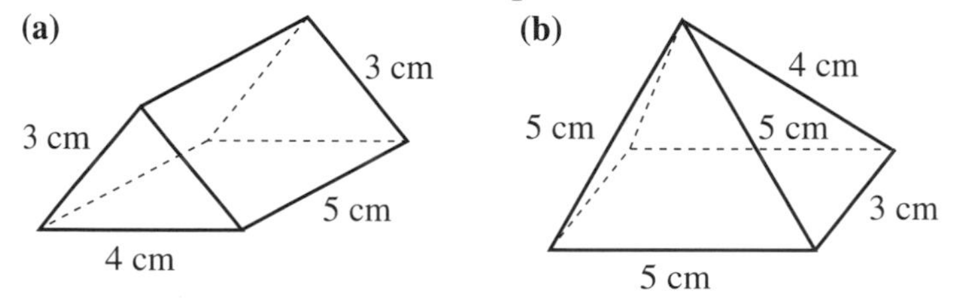

2 Here are some nets of 3-D solids. For each one name the solid and draw a sketch of the solid.

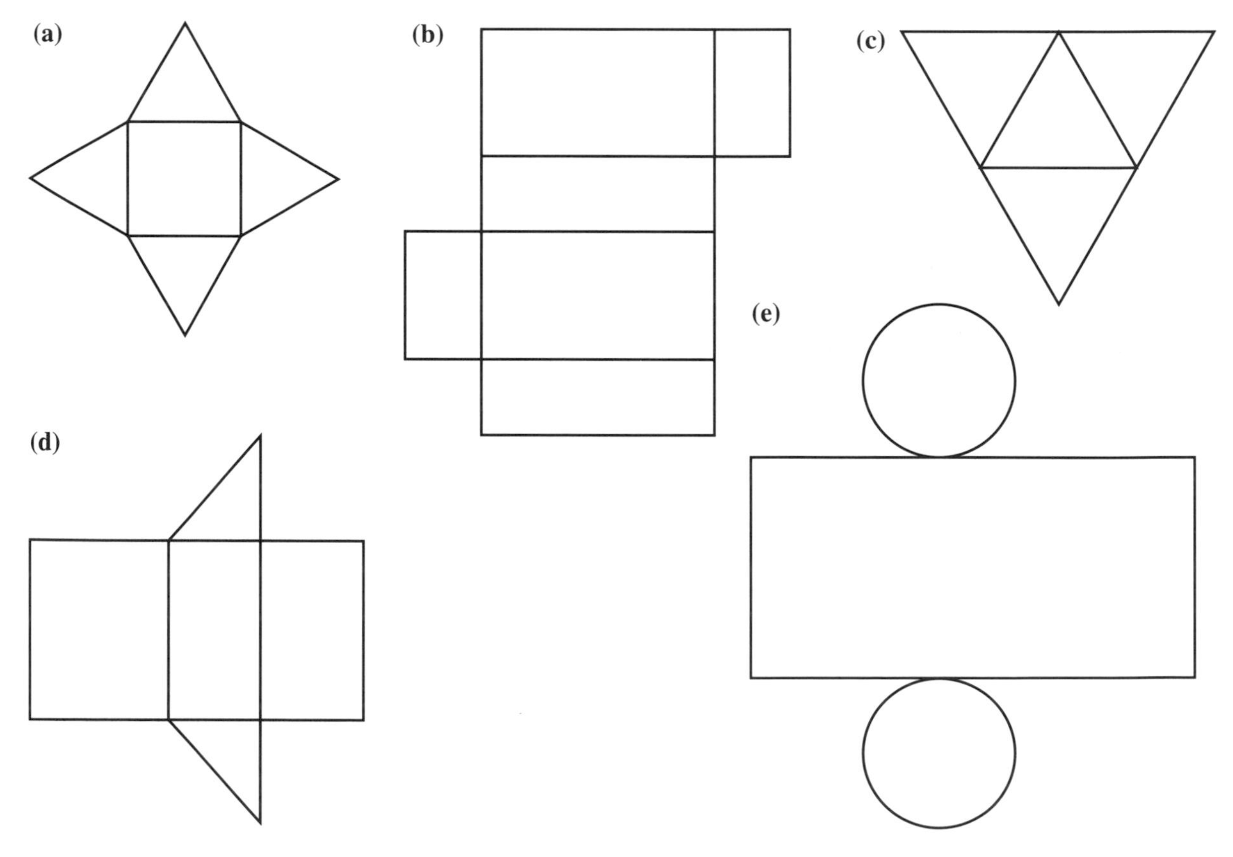

Exercise 4.5 Links: (*4F*) 4F

1 Draw the plan, front and side elevation for these shapes:

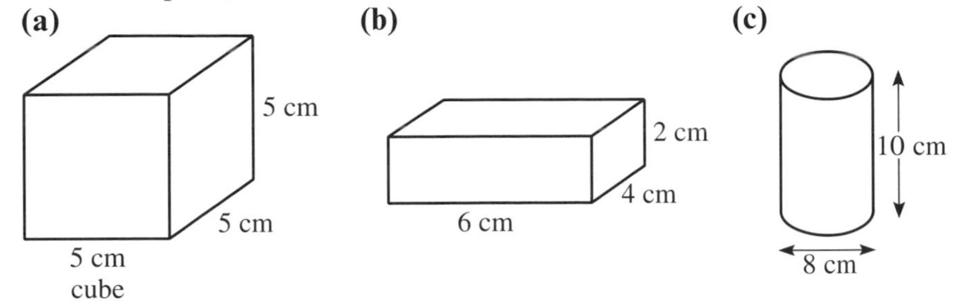

2 Draw the plan, front and side elevation for this shape.

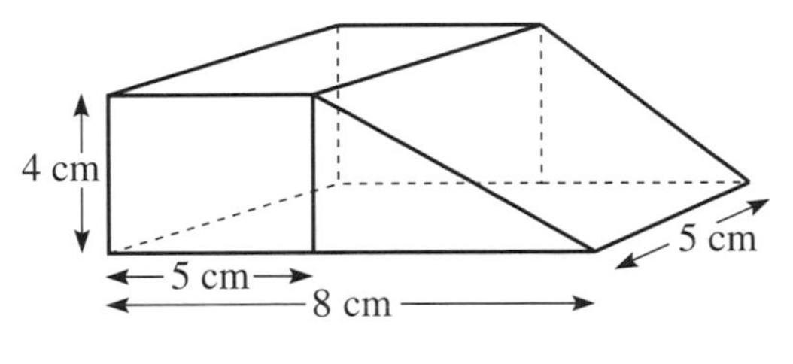

3

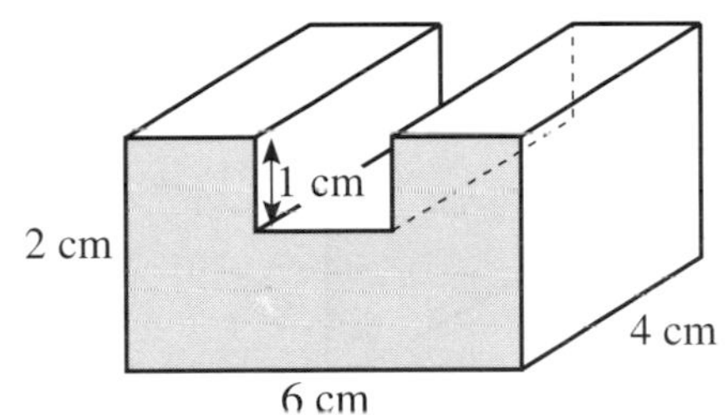

4

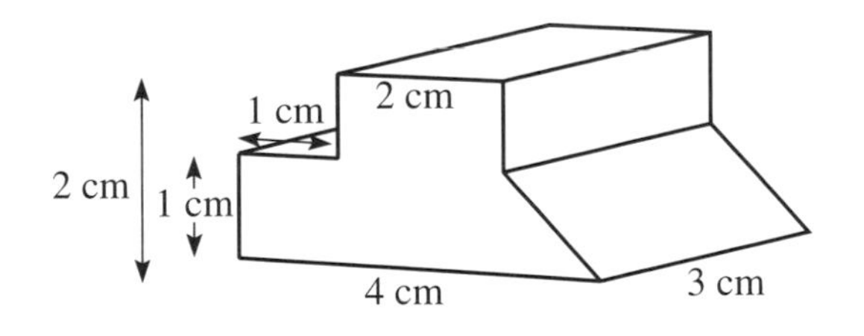

5 Here are the plan, front and side elevation for a shape. Sketch the 3D shape.

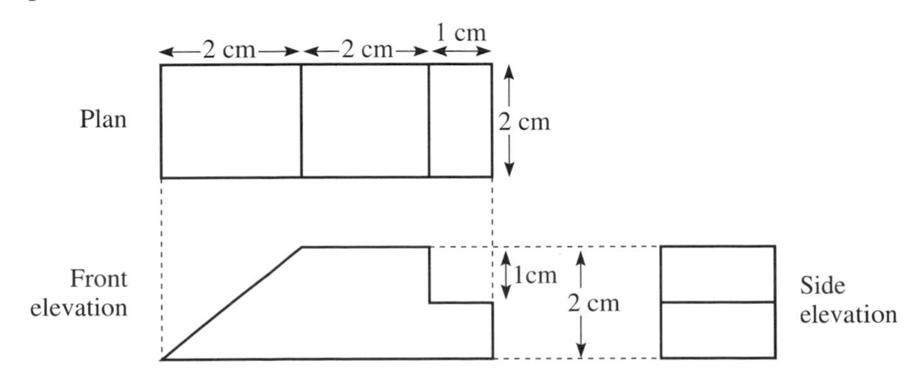

6 Here are the plan, front and side elevation for a shape. Sketch the 3D shape.

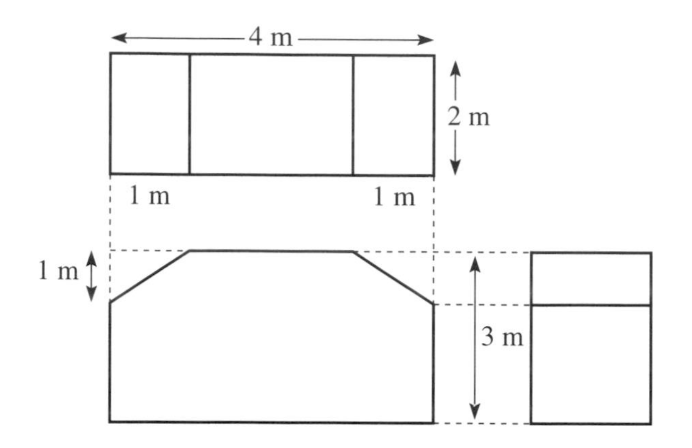

Exercise 4.6 Links: (*4F*) 4G

1 Write down the letters of the pairs of shapes that are congruent.

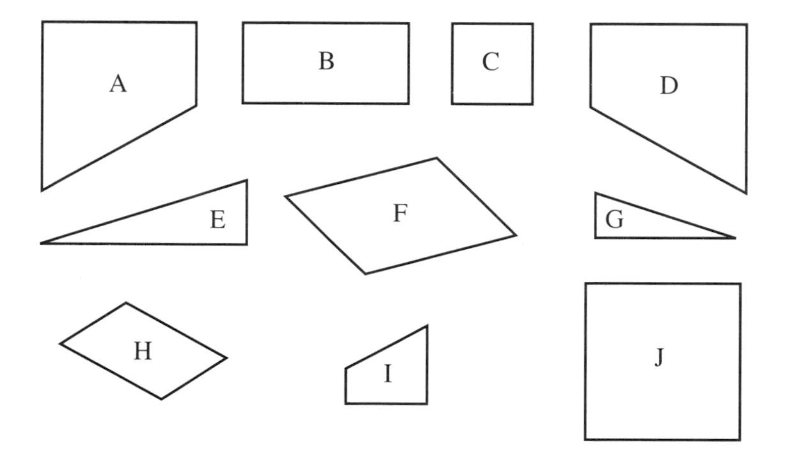

2 Write down the letters of the pairs of shapes that are similar.

3 Copy the shape on to squared paper.
The edge *AB* of a similar shape has been drawn.
Complete the shape.

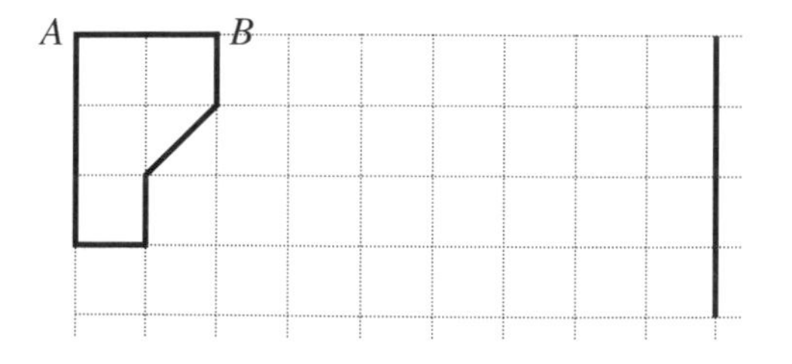

4 **(i)** Draw a shape that has four lines of symmetry.
(ii) Draw a shape that has rotational symmetry of order 3.

5 The two triangles are similar. The lengths of the sides of the large triangle are twice as long as the lengths of the sides of the small triangle.
(i) Work out the size of the angle p.
(ii) Work out the size of the side q.

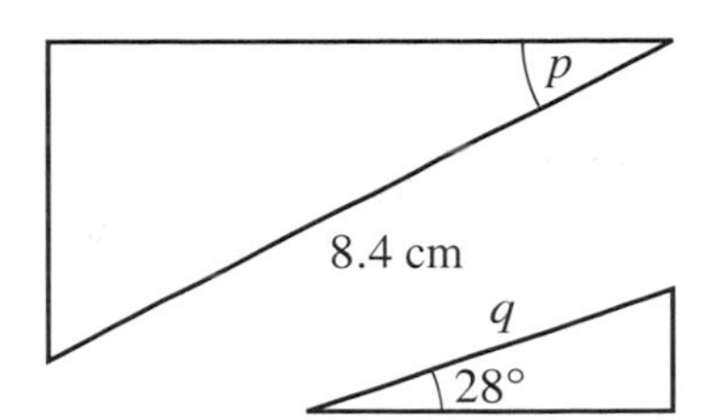

6 Draw a net of a triangular prism.

7 Draw a copy of the shape and show a plane of symmetry for the shape.

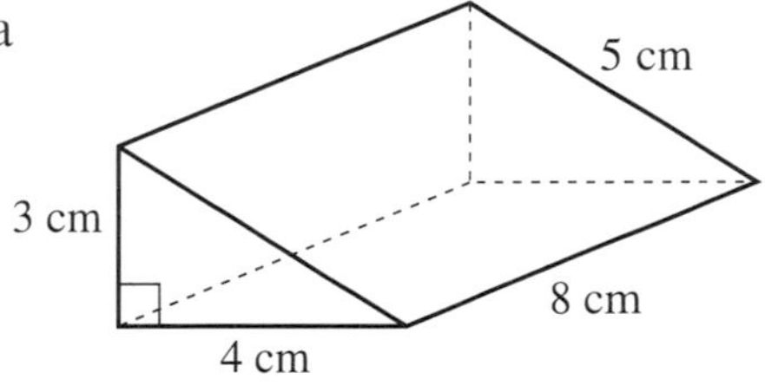

5 Measure 1

Exercise 5.1 Links: (*5A*) 5A

Look at this picture, then write down an estimate for each of the following.

1 The height of the bus
2 The width of the bus
3 The height of the litter bin
4 The height of the woman
5 The height of the lamppost
6 The diameter of the bicycle wheel

7 The length of the car
8 The height of the car
9 The diameter of the wheel
10 The height of the shop window from the ground
11 The width of the shop window
12 The width of the car

Exercise 5.2 Links: (*5B*) 5B

Write down an estimate of the amount of liquid in each of these containers. Give your answers in metric and Imperial units.

1 The coffee cup
2 The coffee maker
3 The mug of tea
4 The glass of milk with a straw in it
5 The wine glass
6 The bottle of champagne
7 The jug when it is full
8 The glass next to the jug

Exercise 5.3 Links: (5*C*) 5C

Write down an estimate of the weight of the following fruits. Give your answer in metric and Imperial units.

1 The 4 bananas.

2 The 3 kiwi fruit.

3 The raspberries.

4 The 5 pears.

5 The pineapple.

6 The bowl of blackberries.

7 A large bag of potatoes.

8 A cabbage.

9 A bag of onions.

10 A box of apples.

11 A basket of potatoes.

12 A melon.

Exercise 5.4 Links: (5*D*) 5D

For each of these statements say whether the measurements are sensible or not. If the statement is not sensible then give a reasonable estimate for the measurement.

1 My mother is 1.60 m tall.

2 My 15 year old brother is 2 cm tall.

3 The front door of my house is 20 m high.

4 My bedroom measures 30 m by 20 m.

5 An apple weighs 2 kg.

6 A can of cola holds about 400 ml of liquid.

7 This page is about 27 mm long.

8 A loaf of bread weighs about 800 g.

9 A box of chocolates weighs about 500 kg.

10 This book weighs about 10 kg.

Exercise 5.5 Links: (*5E*) 5E

Copy and complete this table with appropriate units for each measurement. Give both metric and Imperial units of measurement.

	Metric	Imperial
1 Your height.		
2 The length of your desk.		
3 The weight of a packet of chocolates.		
4 The weight of a large bag of potatoes.		
5 Your weight.		
6 The amount of water in a water barrel.		
7 The amount of liquid in a petrol tanker.		
8 The amount of wine in a wine bottle.		
9 The time it takes to walk one mile.		
10 The time it takes to run 100 metres.		

Exercise 5.6 Links: (*5F – J*) 5F – J

1 Bill arrives at the bus station and looks at his watch. The next bus is due to arrive at 5:30 pm. How long does Bill have to wait until the bus should arrive?

2 Shamus arrives at the train station and looks at the time shown on the station clock. His train is due to arrive at 10:35 am.

(a) How long does he have to wait until the train should arrive?

The train arrives 6 minutes late.

(b) How long did Sheamus have to wait for his train?

3 Change these times from 12-hour clock times (am or pm) to 24-hour clock times.

(a) 9:00 am **(b)** 9:00 pm

(c) 6:30 am **(d)** 6:45 pm

(e) 1 am **(f)** 2 pm

(g) a quarter to eight in the morning

(h) a quarter past seven in the evening

4 Change these times from 24-hour clock times to 12-hour clock times (am or pm).

(a) 07:00 **(b)** 17:00 **(c)** 15:30 **(d)** 08:50
(e) 18:50 **(f)** 07:30 **(g)** 00:10 **(h)** 23:45

5 Write down the cost of posting these letters by **(i)** first class **(ii)** second class.

(a) **(b)**

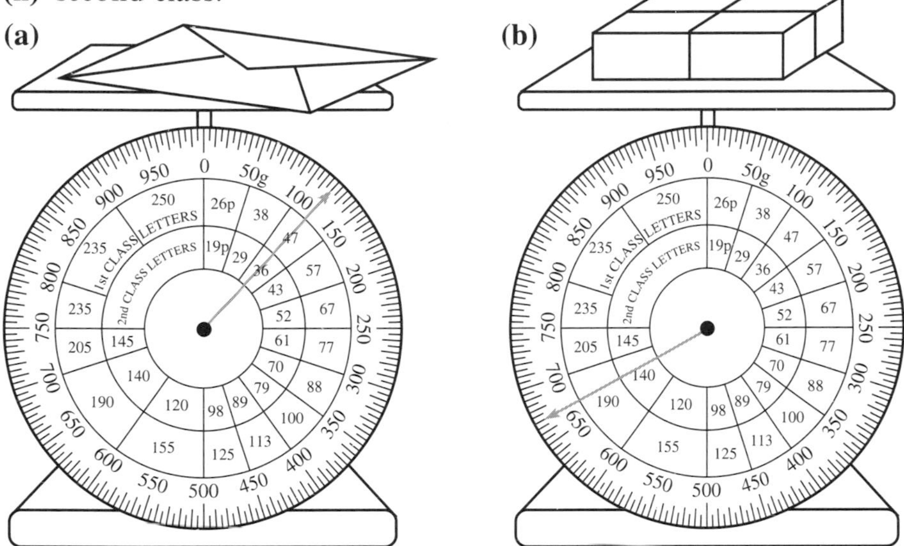

6 Write down the readings on these scales

(a) **(b)** **(c)** **(d)**

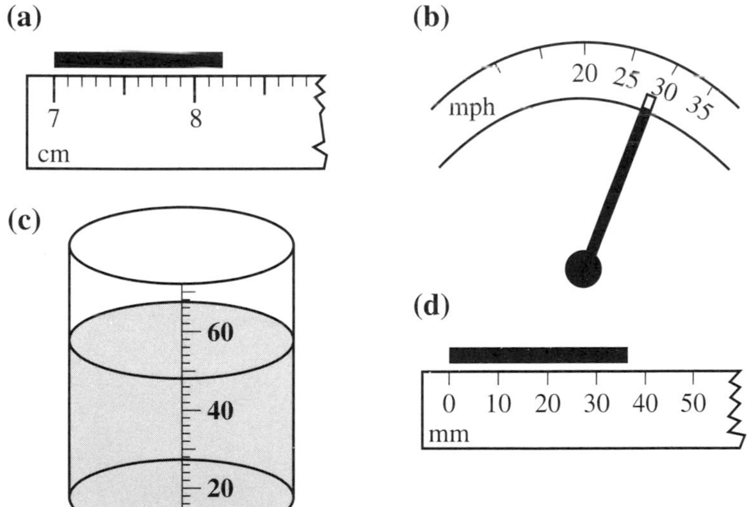

7 Measure and write down the lengths of these lines in centimetres.

(a) ____________________________________

(b) ____

(c) _________________________

(d) __________

(e) ____

8 Draw and label lines of length:
(a) 6 cm **(b)** 5.5 cm **(c)** 3.8 cm **(d)** 10.3 cm
(e) 23 mm **(f)** 55 mm **(g)** 30 mm **(h)** 112 mm

Exercise 5.7 Links: (*5K*, *L*) 5K, L

1 Change these times into minutes:
(a) 3 hours **(b)** 4 hours **(c)** $2\frac{1}{2}$ hours **(d)** $4\frac{1}{4}$ hours
(e) $\frac{1}{2}$ hour **(f)** $\frac{3}{4}$ hour **(g)** 3 hours 35 minutes

2 Change these times into hours:
(a) 120 minutes **(b)** 4 days **(c)** 30 minutes
(d) 300 minutes **(e)** $2\frac{1}{2}$ days **(f)** 200 minutes
(g) 15 minutes **(h)** 2 days **(i)** 150 minutes

3 How many seconds are there in:
(a) 5 minutes **(b)** $\frac{1}{2}$ minute **(c)** $\frac{1}{4}$ minute
(d) 15 minutes **(e)** 1 hour **(f)** $2\frac{1}{2}$ hours?

4 Add 20 minutes to each of these times.
(a) 09:00 **(b)** 10:20 **(c)** 11:45 **(d)** 12:50

5 Add 45 minutes to each of these times:
(a) 09:00 **(b)** 10:20 **(c)** 11:45 **(d)** 12:50

6 Add 3 hours 20 minutes to each of these times:
(a) 10:00 **(b)** 12:30 **(c)** 15:40 **(d)** 16:55

7 Subtract 10 minutes from each of these times:
(a) 10:40 **(b)** 12:00 **(c)** 10:10 **(d)** 09:05

8 Subtract 45 minutes from each of these times:
(a) 13:55 **(b)** 16:45 **(c)** 10:30 **(d)** 08:15

9 Subtract 2 hours 30 minutes from each of these times:
(a) 15:50 **(b)** 16:30 **(c)** 14:10 **(d)** 07:05

10 Alice arrives home at 16:30. She watches television for $2\frac{1}{4}$ hours then spends 1 hour 40 minutes on her homework.
(a) At what time does Alice start her homework?
(b) At what time does Alice finish her homework?

Exercise 5.8 Links: (5*M*) 5M

Use this part of a calendar to answer these questions.

Day	June					July				
Sunday		5	12	19	26		3	10	17	24	31
Monday		6	13	20	27		4	11	18	25	
Tuesday		7	14	21	28		5	12	19	26	
Wednesday	1	8	15	22	29		6	13	20	27	
Thursday	2	9	16	23	30		7	14	21	28	
Friday	3	10	17	24		1	8	15	22	29	
Saturday	4	11	18	25		2	9	16	23	30	

1 What day of the week is the 16th June?
2 Which day and date is 3 days after the 3rd of June?
3 Which day and date is 10 days before the 7th of July?
4 What is the day and date 14 days after the 20th June?
5 What is the day and date a week before the 2nd July?
6 What is the date 2 weeks after the 17th June?
7 What is the date 10 days after the 31st July?

8 Count on 8 days from the following dates.
(a) 4th Jan **(b)** 3rd Feb **(c)** 7th May
(d) 25th Aug **(e)** 25th Sept **(f)** 25th Dec

9 Count back 15 days from the following dates.
(a) 21st Oct **(b)** 16th July **(c)** 29th Feb
(d) 8th July **(e)** 7th May **(f)** 6th Jan

Remember:
30 days hath September
April June and November.
All the rest have 31
except for February alone
which has just 28 days clear
and 29 in each leap year.

Exercise 5.9 Links: (5*N*) 5N

Bus timetable

Bus station	08:00	09:15	10:30
Stadium	08:15	09:30	10:45
High St	08:35	09:50	11:05
Hospital	08:45	10:00	11:15
Museum	08:50	10:05	11:20
Bus station	09:10	10:25	11:40

Train timetable

Swindon	08:00	09:30	12:45
Kemble	08:15	09:45	13:00
Stroud	08:28	09:58	13:13
Stonehouse	08:40	10:10	13:25
Gloucester	08:55	10:25	13:40
Cheltenham	09:05	10:35	13:50

Use the two timetables above to answer these questions.

1 At what time should the 08:00 bus from the bus station arrive at the High St?

2 At what time should the 08:00 train from Swindon arrive in Gloucester?

3 At what time does the 10:05 bus from the museum leave the stadium?

4 At what time does the 9:58 train from Stroud leave Kemble?

5 Buses from the bus station leave every 1 hour 15 minutes. Continue the bus timetable for the next 3 buses. You may assume that each bus takes the same amount of time between stops as the 08:00 bus.

6 The next 2 trains from Swindon leave at 14:15 and 16:05. Continue the train timetable for these next 2 trains. You may assume that each train takes the same amount of time between stops as the 08:00 train.

7 Steven arrives at the train station in Kemble at 09:40. What time is the next train he could catch to Stroud?

8 Florence arrives at the bus stop in the High St at 11:00. What time is the next bus she could catch to the Museum?

9 Work out how long it should take to travel between:
(a) the High St and the museum
(b) the hospital and the bus station
(c) the stadium and the museum
(d) Kemble and Stonehouse
(e) Stroud and Gloucester.

10 The bus timetable is for a bus route in Gloucester. It takes seven minutes to walk from the bus station to the train platform. Use both timetables to work out which bus and train:
(a) Henry catches at the museum to be in Cheltenham by 10:45
(b) Phil catches at the hospital to be in Cheltenham by 14:00
(c) Sue catches at the stadium to be in Cheltenham for 17:30.

11 Jim got up at 07:40. He took 35 minutes to wash, dress and have breakfast. It took him 10 minutes to walk to the bus stop. The bus arrived 7 minutes later. Jim arrived at work 35 minutes later.
(i) At what time did Jim arrive at the bus stop?
(ii) At what time did Jim arrive at his destination?

Jim was due to start work at 08:30.
(iii) How late was he for work?

The next day Jim was exactly on time for work.
All the amounts of time remained the same.
(iv) At what time did Jim get up?

6 Approximation

Exercise 6.1 — Links: (*6A*, *B*) 6A, B

1 Round these to:
(i) the nearest 10 **(ii)** the nearest 100 **(iii)** the nearest 1000

(a) 4970	**(b)** 7456	**(c)** 9011	**(d)** 1649
(e) 8648.5 m	**(f)** 9999.9 km	**(g)** £26 546	**(h)** £2957.59

2 The attendance at a soccer match was 18 347. Write this figure to the nearest 1000.

3 The quotation for a building extension is £26 499. Write this quotation to the nearest £1000.

4 Cola cans come in packs of 8. How many packs are needed to have 76 cans?

5 Chocolate biscuits are sold in packs of 6. Freda needs 40 chocolate biscuits. How many packs does she need to buy?

6 Wood screws are sold in packs of 25. Richard needs 180 wood screws. How many packs should he buy?

7 A coach can carry 53 passengers. 255 people need to travel. How many coaches are needed?

8 Wine is packed in crates of 12 bottles. 928 bottles are to be packed. How many crates are needed?

Exercise 6.2 — Links: (*6C*, *D*) 6C, D

1 Round each number to **(i)** 3 d.p. **(ii)** 2 d.p.

(a) 0.0467	**(b)** 17.3695	**(c)** 8.0459	**(d)** 15.1064
(e) 13.5941	**(f)** 0.0067	**(g)** 76.9750	**(h)** 4.9999
(i) 23.0106	**(j)** 19.3469	**(k)** 1.3094	**(l)** 18.9766
(m) 7.3949	**(n)** 10.9876	**(o)** 10.9816	**(p)** 43.9954

2 Round each number to **(i)** 1 s.f. **(ii)** 2 s.f. **(iii)** 3 s.f.

(a) 2376	**(b)** 0.035 92	**(c)** 249.5	**(d)** 167.5927
(e) 0.099 95	**(f)** 3.4675	**(g)** 9.435	**(h)** 76.75
(i) 19.45	**(j)** 11.93	**(k)** 45 647	**(l)** 1546
(m) 1999.99	**(n)** 77.471	**(o)** 0.5498	**(p)** 751.67

Exercise 6.3 Links: (*6E*) 6E

1 **(i)** Write down and use a suitable calculation to check each answer shown.
(ii) If the answer shown is incorrect, work out the correct answer to the original problem.

(a) $378 \times 9 = 3402$ **(b)** $3432 \div 8 = 429$
(c) $623 + 489 = 1112$ **(d)** $4063 - 94 = 969$
(e) $6.521 \div 2.003 = 3.526$ **(f)** $109.8091 - 98.9 = 8.919$

2 **(i)** Write down a calculation that could be used to estimate the answer.
(ii) Work out the estimated answer.
(iii) Use a calculator to work out the exact answer.

(a) 4.78×9.17 **(b)** $\dfrac{78.3 \times 8.1}{0.23}$
(c) $18.46 \div 4.71$ **(d)** $(99.7 \times 509) + 349$
(e) $\dfrac{0.0354 \times 0.496}{0.054}$ **(f)** $\dfrac{0.789 \times 7.225}{4.96 \times 0.291}$

Exercise 6.4 Links: (*6F, G*) 6F, G

For each measurement in questions **1** to **5** write down:
(i) the minimum it could be **(ii)** the maximum it could be.

1 These measurements are given correct to the nearest whole number:
(a) 54 g **(b)** 17 cm **(c)** 375 *l* **(d)** 50 kg

2 These measurements are given correct to one decimal place:
(a) 98.7 c*l* **(b)** 19.2 g **(c)** 46.1 m **(d)** 64.4 cm

3 These measurements are given correct to two decimal places:
(a) 8.36 mm **(b)** 15.09 c*l* **(c)** 0.57 g **(d)** 13.48 s

4 These measurements are given correct to three decimal places:
(a) 1.456 g **(b)** 4.959 s **(c)** 0.545 *l* **(d)** 1.985 g

5 Write 8528 to the nearest 10, 100 and 1000.

6 A farmer has 900 cabbages to send to market. They are packed in crates of 24. How many crates are needed?

7 Round these numbers to 2 decimal places and 2 significant figures:
(a) 19.261 **(b)** 1097.254 **(c)** 0.012 69 **(d)** 5.3546

8 Work out an estimate for the calculation:

$$\frac{278 - 109.2}{8.4}$$

Write down the numbers you used to estimate this calculation.

9 The distance from Bedford to Clifton is given on a signpost as 11 miles.
(a) What is the maximum distance it could be?
(b) What is the minimum distance it could be?

10 **(a)** Write down the numbers you could use to get an approximate answer to

$$\frac{87 \times 41}{18}$$

(b) Write down your approximate answer. Do not use a calculator.
(c) Work out the exact answer to 2 decimal places.
(d) Work out the difference between your approximate answer and the exact answer.

11 The distance between Manchester and Chorley is measured as 35.7 miles.
(a) Write down the minimum distance it could be.
(b) Write down the maximum distance it could be.

12 Freda wants to estimate the value

217×47.8

Each number must be written to 1 significant figure.
(a) Write down a suitable calculation which could be used.
(b) Write down the value of the estimate.

7 Graphs of linear functions

Exercise 7.1 Links: (*7A, B*) 7A, B

1 **(a)** Write each of these functions as an equation using x and y.

(i) $n \rightarrow n + 3$ **(ii)** $n \rightarrow n - 1$
(iii) $n \rightarrow 3n$ **(iv)** $n \rightarrow 3n + 4$

(b) Make a table of values for each equation using the values 0, 1, 2, 3, 4, 5 for x.

(c) Draw the graphs to show the numbers for each of your tables of values.

2 **(a)** Copy and complete the table of values for this sequence of dot patterns:

Stage x	1	2	3	4	5
Number of dots y	5	8	11		

(b) Draw the graph for this table of values.
(c) Write down the equation of this graph.

3 **(a)** Invent a sequence of matchstick patterns for the table of values below:

x	1	2	3	4	5
y	6	11	16	21	26

(b) Draw the graph for this table of values.
(c) Find the equation of this graph.

4 Two common Imperial units of measure are yards and feet. They are connected by the rule

1 yard = 3 feet

(a) Copy and complete this table of values:

Yards x	0	1	2	3	4	5	6
Feet y	0	3				15	

(b) Draw the graph for this table of values.
(c) Write down the equation of this graph.
(d) By using your graph, or otherwise, find:
(i) the number of feet equal to 2.5 yards
(ii) the number of yards equal to 10 feet.

5 Karen saw an advertisement for Lucea Aerobics Club. The advertisement read:

You pay a £60 joining fee then you pay £2 for each session you attend.

(a) Work out how much it would cost Karen to join the club and then attend:
(i) 10 sessions **(ii)** 50 sessions **(iii)** 100 sessions

(b) Using x to stand for the number of sessions and y to stand for the cost (in pounds), draw the graph to show the cost of joining the club and then attending any number of sessions up to and including 100.

(c) Work out the equation of this graph.

Exercise 7.2 Links: (*7C*) 7C

1 A car travels 32 miles on each gallon of petrol.

(a) Draw a graph to show how many miles the car will travel on any amount of petrol up to and including 10 gallons.

(b) Use your graph to find:
(i) the distance the car will travel on 4.5 gallons of petrol
(ii) the amount of petrol the car will use on a journey of 200 miles.

2 The cost, £y, of renting a van for a day and driving it x miles is given by the equation $y = 0.4x + 30$

(a) Copy and complete this table of values for x and y:

Miles x	0	50	100	150	200	250	300
£y	30					130	

(b) Draw the graph for this table of values.

Mrs Taylor rents a van for a day. Use your graph to answer these questions.

(c) What is the total cost of renting the van for a day and driving 130 miles?

(d) What was the distance she travelled when the total cost was £120?

3 **(a)** On the same axes, draw and label with the letters, the graphs of:

A $y = 3x$ **B** $y = 3x + 1$ **C** $y = 3x + 4$ **D** $y = 3x - 2$

for values of $x = 0, 1, 2, 3, 4$.

(b) Write down at least three things that **all four** graphs have in common.

4 **(a)** On the same axes, draw and label the graphs:

A $y = x + 2$ **B** $y = 3x + 6$

for values of $x = 0, 1, 2, 3, 4$.

(b) What do you notice about these two graphs?

5 **(a)** Copy and complete this table of values for $y = 12 - x$.

x	0	1	2	3	4
$y = 12 - x$	12				8

(b) Draw the graph of $y = 12 - x$ for values of x from 0 to 4.
(c) On the same axes and for the same values of x, draw the graph of $y = 5 - x$.
(d) What do you notice about the two graphs?

Exercise 7.3 Links: (*7D, E, F*) 7D, E, F

1 **(a)** Copy and complete the table of values for $y = 2x + 5$.

x	−3	−2	−1	0	1	2	3
y		1					11

(b) Draw the graph of $y = 2x + 5$ for values of x from −3 to 3.

2 $y = 3x + 2$
(a) Make a table of values for x and y for all values of x from −4 to 4.
(b) Draw the graph of $y = 3x + 2$ for values of x from −4 to 4.

3 **(a)** Copy and complete the table of values for $y = 10 - 2x$.

x	−3	−2	−1	0	1	2	3
y	16			10			4

(b) Draw the graph of $y = 10 - 2x$ for values of x from −3 to 3.

4 Find the equation of each of these straight lines:
(a)

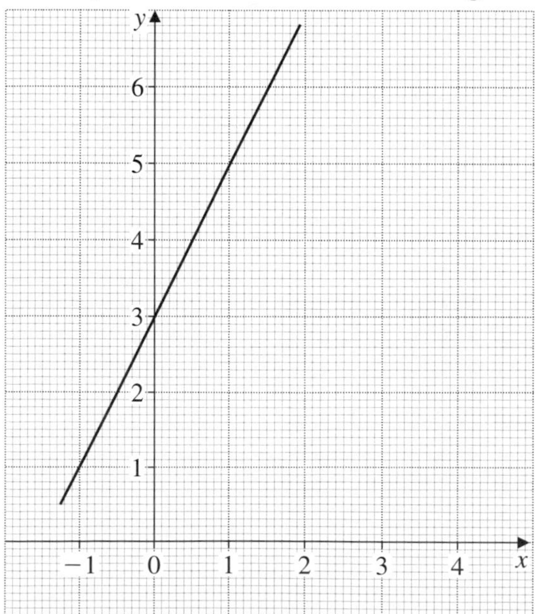

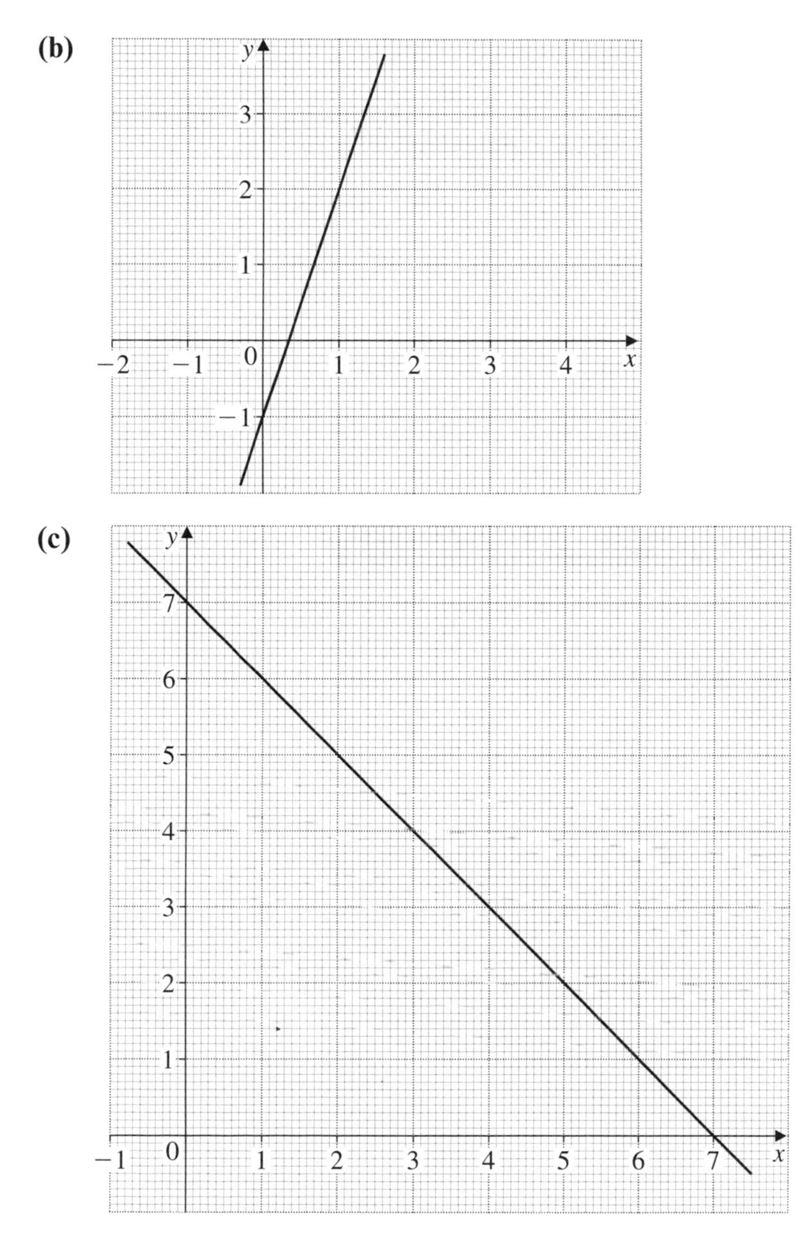

5 **(a)** On graph paper, draw the straight line which passes through the points (0, 2) and (1, 5).

(b) Find the equation of this line.

6 **(a)** Draw the straight line with equation $y = 4x - 5$ for values of x from -2 to 4.

(b) Use your graph to find the value of x when $y = 1$.

7 **(a)** On graph paper, draw the straight line which passes through the points $(-1, -3)$ and (3, 5).

(b) Find the equation of this line.

8 A straight line has a gradient of 3 and passes through the point (0, 1). Write the equation of this line in the form $y = mx + c$.

9 On the same axes draw and label each of these lines:

A $y = 2$ **B** $y = -2$ **C** $y = 5$ **D** $y = 0$

10 **(a)** Draw the line $y = 3$.
(b) On the same axes, draw the line $x = 2$.

These two lines meet at the point P.
(c) Write down the coordinates of the point P.

The coordinates of the point Q are $(0, -1)$.
(d) On the same axes plot and label the point Q.
(e) Draw the line joining P and Q.
(f) Find the equation of the line joining P and Q.

Exercise 7.4 Links: (*7G*) 7G

1 Last Saturday Mrs Ijoma left her home in Redwell and travelled by car to Lucea to do some shopping.
She left home at 9:00 am.
She arrived back home at 5:00 pm.
Her journey from Redwell to Lucea and back is represented by the distance/time graph below.

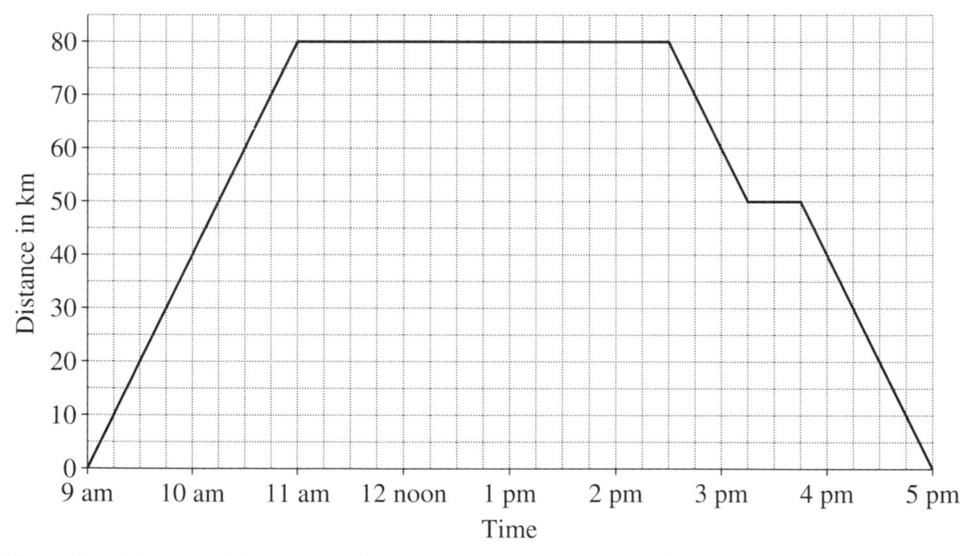

Use the distance/time graph to answer these questions:
(a) How far is it from Redwell to Lucea?
(b) How long did it take Mrs Ijoma to travel from Redwell to Lucea?
(c) How long did Mrs Ijoma spend in Lucea?
(d) At what time did she leave Lucea to return to Redwell?

On the return journey from Lucea back to Redwell Mrs Ijoma stopped for a cup of tea.
(e) At what time did she stop for a cup of tea?
(f) How long did she stop for the cup of tea?
(g) How long did the return journey from Lucea to Redwell take?

On the journey from Redwell to Lucea, Mrs Ijoma drove at a constant speed.

(h) What was that constant speed in miles per hour?

2 Maureen left home at 10:00 am to take her dog Kaytu for a walk. She walked at a constant speed for 15 minutes during which she covered a distance of 1200 metres to the local playing field. At the playing field she sat on a bench whilst Kaytu played in the field for 25 minutes. Maureen then took Kaytu back to her home. She arrived back home at exactly 11:00 am.

(a) Draw a distance/time graph for this situation.

(b) At what time did Maureen and Kaytu leave the playing field?

(c) Work out Maureen's speed in metres per minute for the walk from her home to the playing field.

3

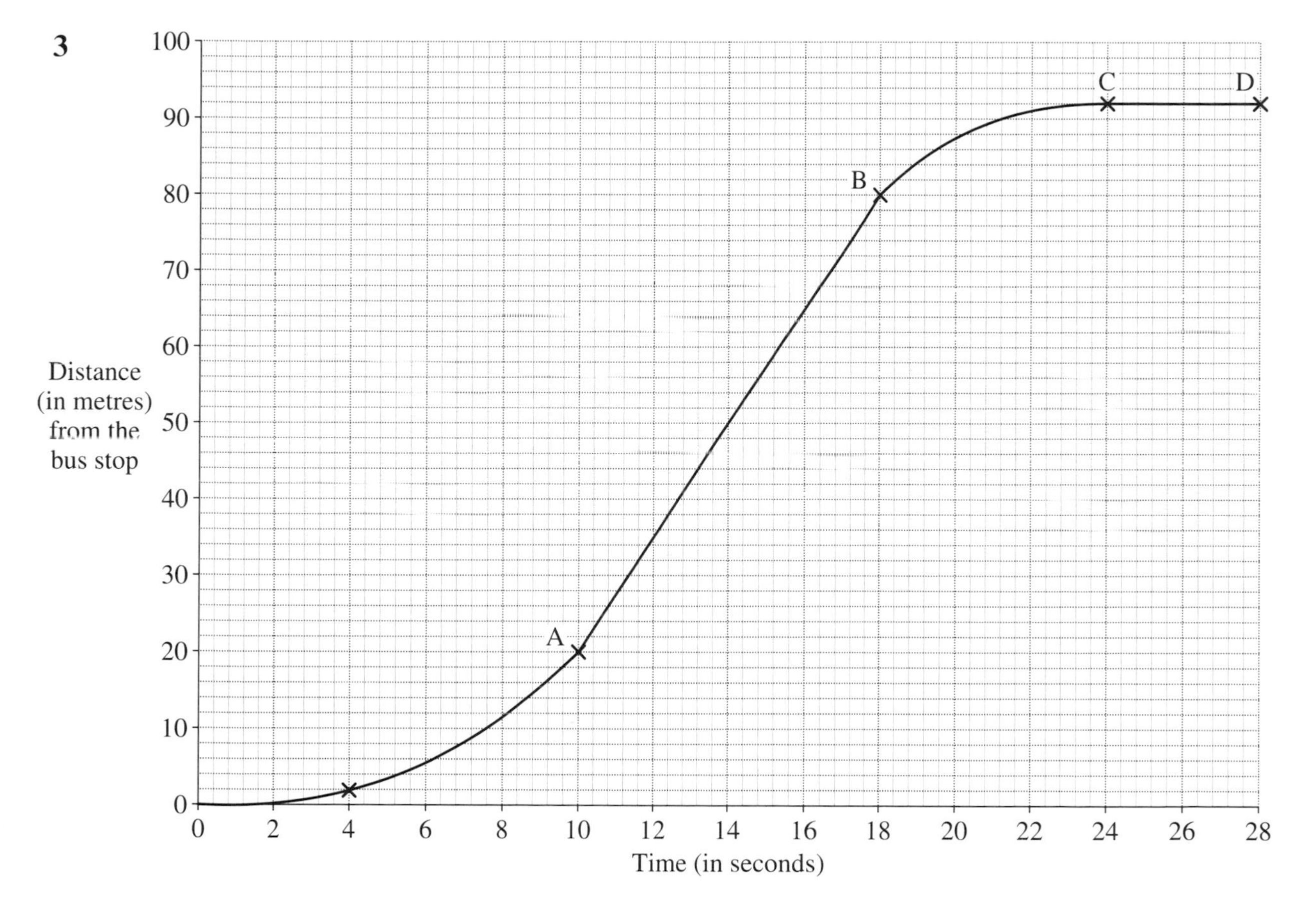

The diagram shows a distance/time graph for a bus after it had left a bus stop.

(a) Use the graph to find the distance the bus travelled in the first 18 seconds after it had left the bus stop.

(b) Describe fully the journey of the bus represented by the parts AB, BC and CD of the graph.

(c) Using the graph, work out the speed of the bus 16 seconds after it had left the bus stop.

8 Collecting and organizing data

Exercise 8.1 Links: (*8A*) 8A

1 The table below contains some of the information about attendance at a rock festival.

	Male	Female	Total
20 and under		14%	24%
21–24	23%	21%	
25 and over			
Totals	57%		

Complete the table.

(a) What was the percentage of females at 25 and over?

(b) What was the percentage of males at 25 and over?

2 A supermarket stocks three brands of washing soap; Washo, Cleana and Brighta. Each brand has powder and liquid versions. Only some of the sales records for the previous week can be found. The overall total sales were 850 of which Brighta contributed 317. The sales of powder included 106 packets of Washo and 89 packets of Cleana. Liquid sales included 221 of Brighta and 151 of Cleana. Construct a two-way table to work out the missing details.

(a) What were the total sales of Washo?

(b) What was the total of packets of powder sold?

3 The supermarket stocks a well-known brand of cornflakes as well as its own brand. Both brands are supplied in 350 g, 500 g and 750 g packets. 350 g packets account for 27% of the sales and 750 g packets for 52%. The store's own brand accounts for one third of the 350 g and one quarter of the 750 g packets sold. Equal numbers of packets of 500 g are sold for each brand. Draw up and complete a two-way table.

(a) What percentage overall are represented by the store's own brand?

(b) The total sales for the week are 1300 packets. How many of these are the store's own 750 g packets?

4 The table shows the supermarket coffee sales.

	100 g	200 g	400 g	Total
Store brand A		126		
Brand B granules	120		53	
Brand B powder	41	86		147
Brand C powder	66		31	193
Total	300		132	

The manager remembers that they did not have any stocks of Brand B 200 g granules to sell. That enables the table to be completed.

(a) What was the total number of jars of coffee sold?

(b) How many 200 g jars of coffee were sold?

5 50 pupils are going on an educational visit. The pupils have to choose to go to one of the theatre, the art gallery or the science museum.
23 of the pupils are boys.
11 of the girls choose to visit the theatre.
9 of the girls choose to visit the art gallery.
13 of the boys choose to visit the science museum.
6 more pupils visit the art gallery than visit the theatre.
Draw up and complete a two-way table.
How many of the girls choose to visit the science museum? [E]

Exercise 8.2 Links: (*8B*) 8B

Design suitable tables for recording the data in each of the following questions.

1 Writing paper is stocked in white, azure, blue and grey. The sizes available are 20 cm × 16 cm, A4, A5. Paper is graded by weight: 80 gsm and 100 gsm are available.

2 Second-hand cars by make, engine size and age.

3 Cardigans come in several sizes, colours and fabrics. Use sizes S, M, L, XL; colours red, green and blue; fabrics wool, nylon, acrylic.

4 Caravan hire to include a range of sites; 2, 4, 6, 8 berth caravans which can be luxury or tourist; and the weekly rate for different periods of the year.

Exercise 8.3 Links: (*8C*) 8C

The table contains information about flowers that are available at a nursery.

Flower	Colours	Growing heights	Uses	Fragrance
Hyacinth	White, blue	15 cm	Window box Hanging basket	Yes
Sunflower	Yellow	90 cm	Cut flower	No
Chrysanthemum	All	30 cm, 60 cm	Cut flower	No
Scarlet Pimpernel	Red, blue	15 cm, 30 cm	Window box Hanging basket	No
Carnation	White, red, multicolour	30 cm, 60 cm	Cut flower	Yes
Dahlia	White, red, yellow	15 cm, 30 cm, 60 cm	Cut flower	No
Browallia	White, blue	30 cm	Window box Hanging basket	Yes
Nasturtium	Yellow, pink, red	30 cm, 60 cm	Cut flower Window box Hanging basket	Yes

1 To complete a window box, Mary needs a small blue plant. What are her choices?

2 Jaspal needs a tall plant to stand at the back of a flower bed. He wants white and red flowers. What can he choose?

3 Kate's hanging basket needs a medium (30 cm) size yellow flower for the centre. What is available?

4 Anne wants to grow fragrant plants that can be used for floral displays in vases. Colour is not important. What can she choose?

5 Victor needs a plant for a window box. His choice has to be medium (30 cm) height and fragrant. What is available?

Exercise 8.4 Links: (*8D, E, F*) 8D–G

1 For each type of data write down whether it is qualitative, quantitative and discrete, or quantitative and continuous:

(a) the length of a TV programme
(b) the marks in a test
(c) the height of a mountain
(d) the result of a game
(e) the feel of a surface
(f) the volume of a sphere
(g) the thickness of custard
(h) the capacity of a box
(i) the age, to the nearest year, of a building

2 Robin had a holiday job packing cheese. Each pack of cheese should weigh 500 grams. Robin had to pack 30 packs of cheese. Robin checked the weights, in grams, correct to the nearest gram. These are the results:

512 506 503 506 499 499 500 504 502 503
496 497 497 509 506 499 497 498 507 511
498 491 496 506 507 493 496 503 510 508

Copy and complete the grouped frequency table for the weights. Use class intervals of 5 g.

Weight (w grams)	Tally	Frequency
$490 \leqslant w < 495$		

[E]

3 Alan is doing a survey of the heights of boys and girls in Year 7. He first takes a random sample of 70 boys from Year 7. Suggest a suitable method that Alan could use to take a random sample. [E]

4 Data on the distance and bearing (but not the altitude) of aircraft in the vicinity of Heathrow is taken at noon one day in the peak summer season. This data is given below in coordinate form (distance, bearing).

(5, 123) (7, 080) (14, 130) (9, 275) (11, 189) (3, 021)
(2, 030) (1, 039) (6, 147) (8, 152) (14, 190) (17, 005)
(23, 320) (11, 216) (12, 234) (30, 300) (18, 010) (6, 095)
(4, 130) (8, 235) (11, 175) (6, 125) (9, 300) (16, 165)
(12, 060) (14, 130) (1, 220) (7, 215) (11, 155) (20, 081)
(13, 110) (17, 215) (14, 280) (4, 290)

Construct and complete a two-way table with the distance categories $0 < x < 5$, $5 \leqslant x < 10$, $10 \leqslant x < 15$, $x \geqslant 15$ and the bearings as the four quadrants.
Which is the most crowded region?

Exercise 8.5 Links: (*8G*) 8H – K

1 Karl's and Eleanor's school is near a busy main road. They decide to carry out a survey of the different types of vehicle that travel on the main road.
Design a suitable data sheet so that they can collect their data easily. [E]

2 Fred is conducting a survey into television viewing habits. One of the questions in his survey is:

'How much television do you watch?'

His friend Sheila tells him that it is not a very good question. Write down two ways in which Fred could improve his question. [E]

3 Martin, the local Youth Centre leader, wishes to know why attendance at the Youth Centre is less than at the same time last year. He thinks that it could be due to a number of changes that occurred during the course of the year. These changes were:

the opening hours changed
a new sports centre opened nearby
some of the older members started bullying the younger members

Design a suitable question, that is easily answered, to find out why people do not attend the Youth Centre. [E]

Note: There are no Practice exercises for Unit 9: Using and applying mathematics.

10 Angles

Exercise 10.1 Links: (*10A, B*) 10A, B

1 Work out the lettered angles in these shapes. Write down your reasons.

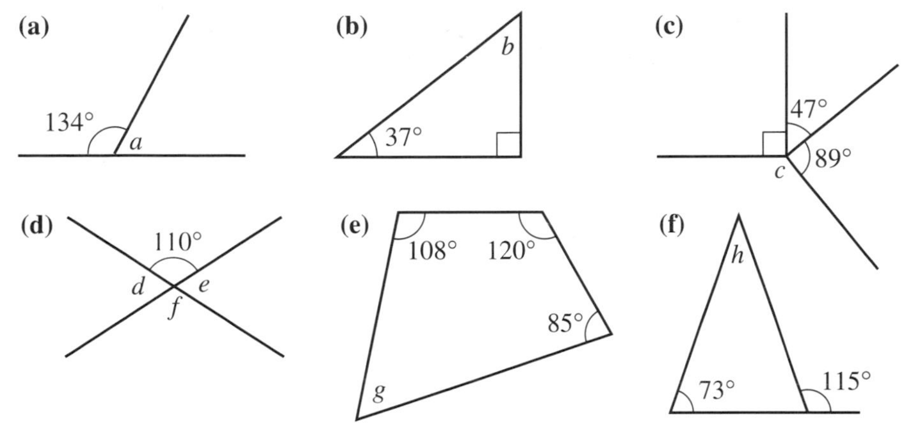

2 Draw a copy of each diagram and find all the unmarked angles. Write down the reasons.

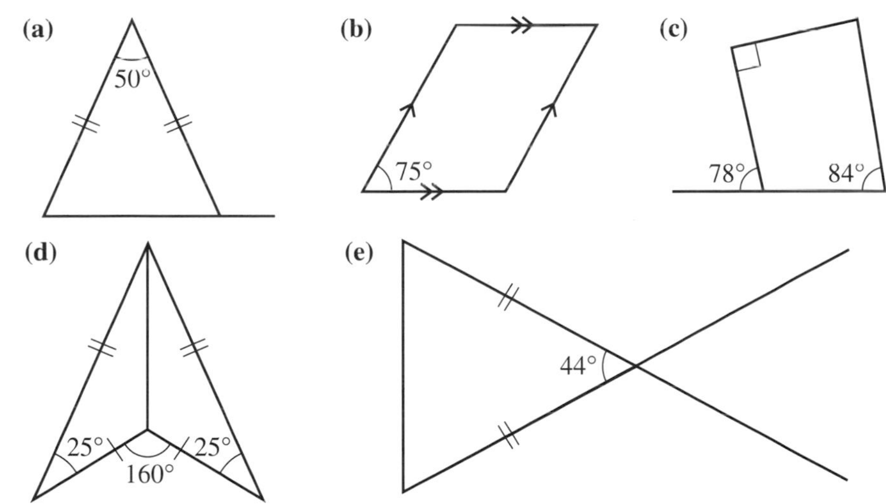

3 Calculate the size of the interior angle of a regular hexagon.

4 Calculate the unknown angles in these polygons.

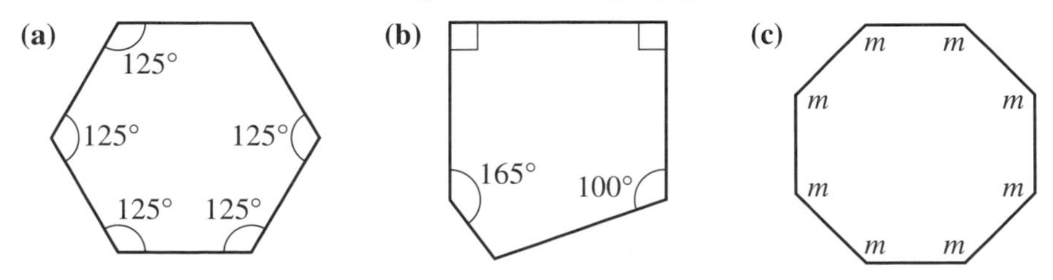

5 This diagram shows a regular octagon.
Calculate the value of x and y.

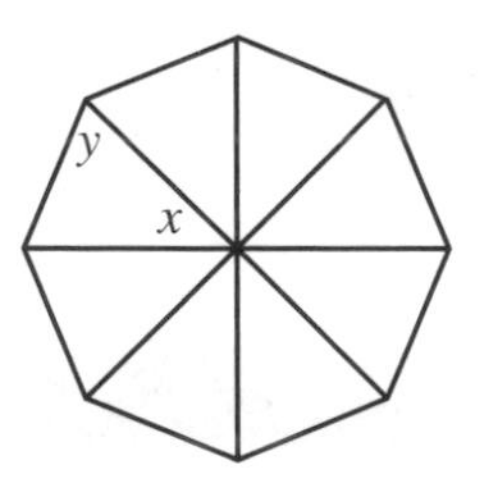

Exercise 10.2 Links: (*10C – E*) 10C – E

1 Calculate the named angles, giving reasons.

(a)

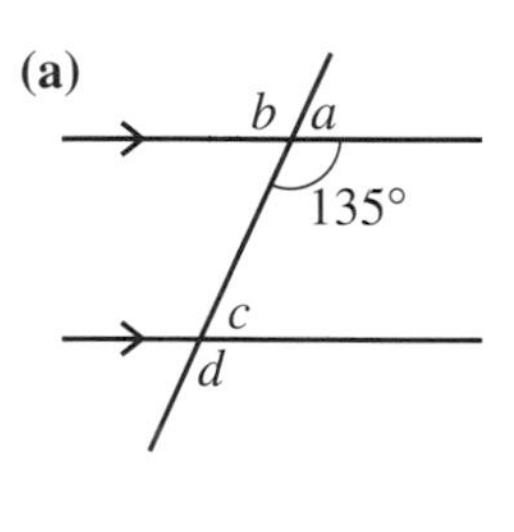

(b)

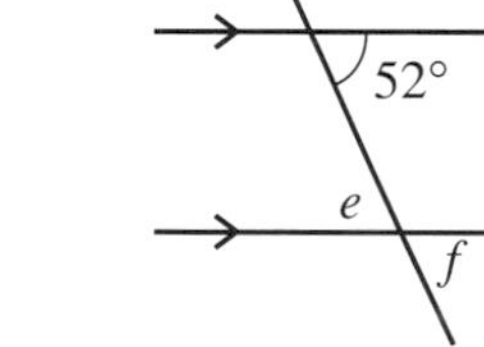

(c)

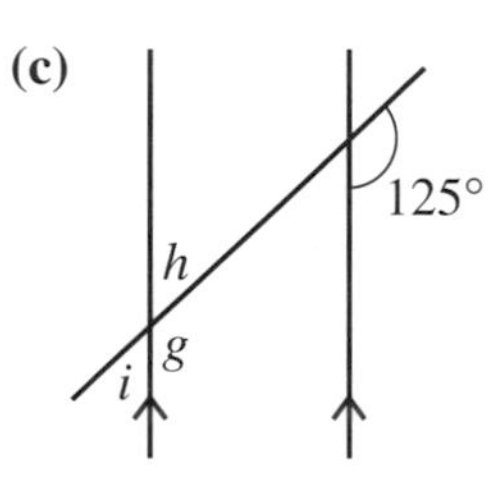

(d)

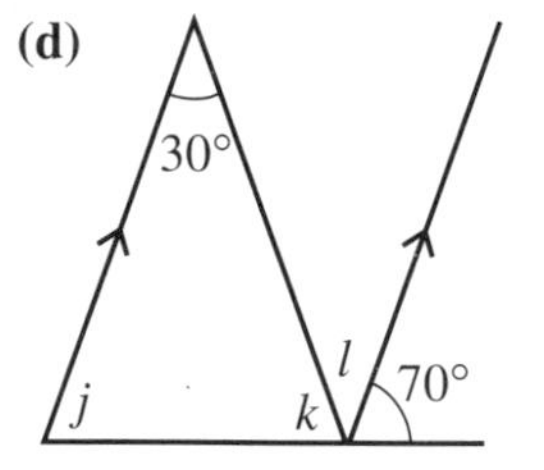

(e)

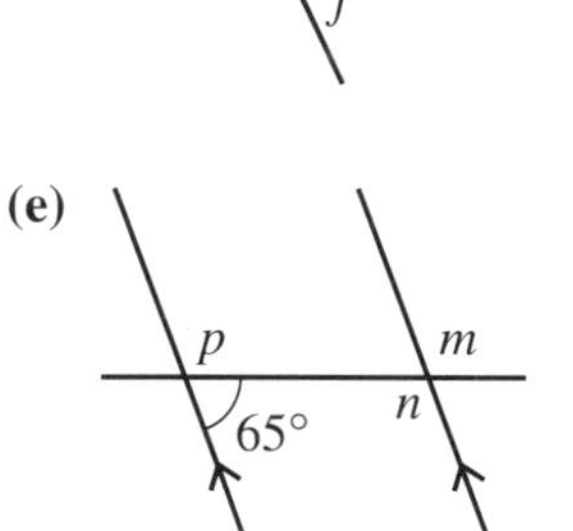

(f)

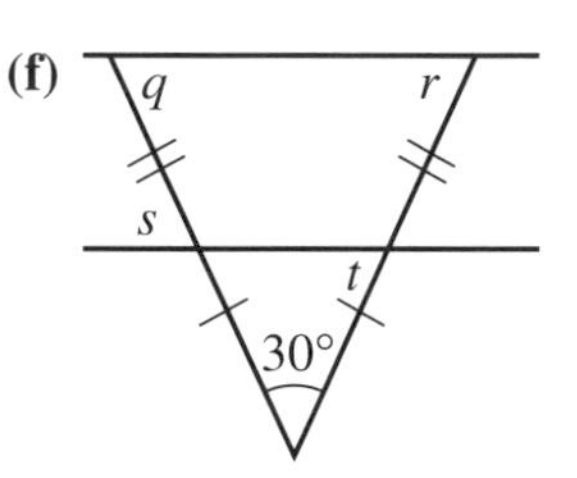

2 Calculate the named angles, giving reasons.

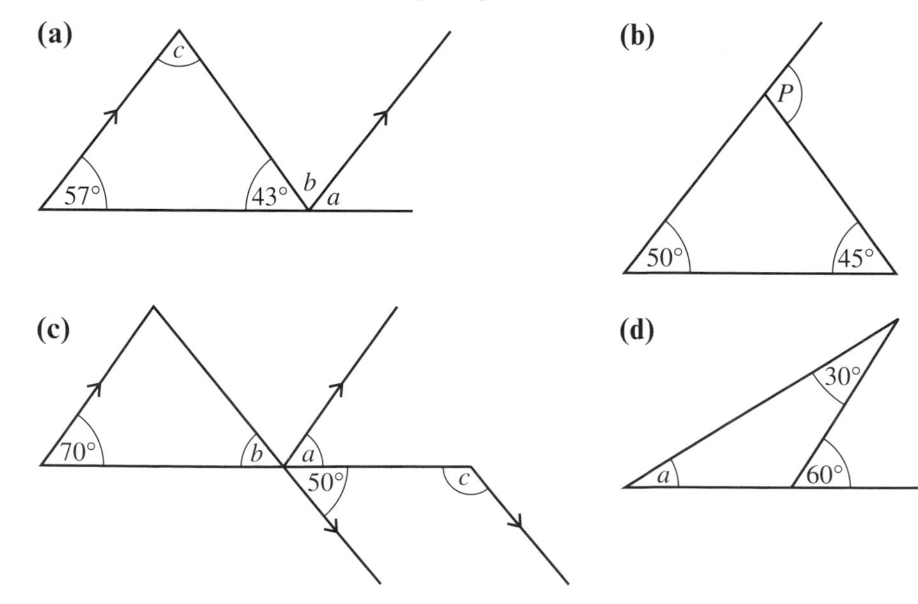

3 Draw two parallel lines cut by another straight line.

(a) Mark with the letter A a pair of alternate angles.

(b) Mark with the letter C a pair of corresponding angles.

(c) Mark with the letter V a pair of vertically opposite angles.

4 Calculate the marked angles in these diagrams. O is always the centre of the circle.

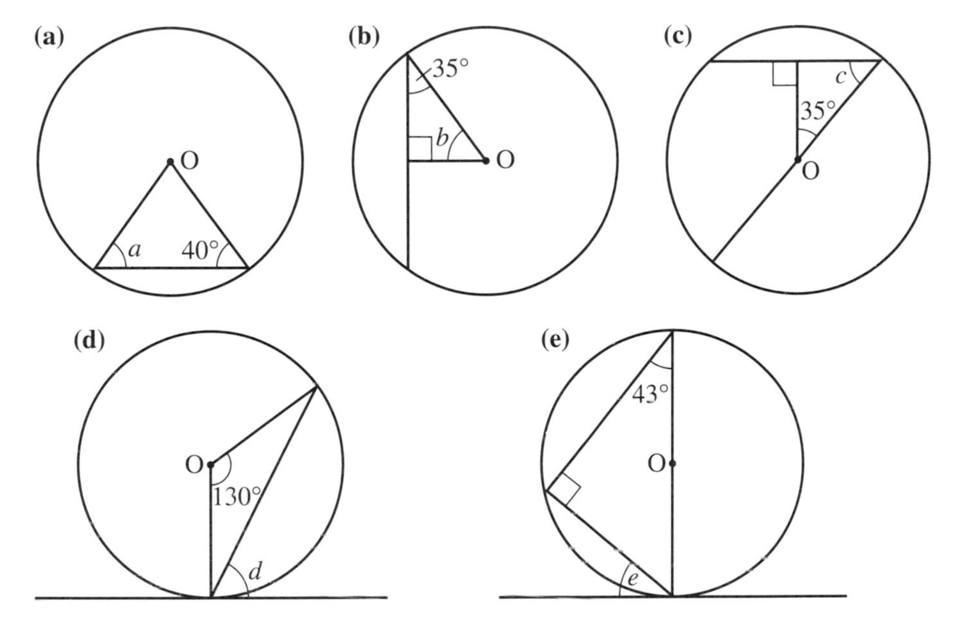

5

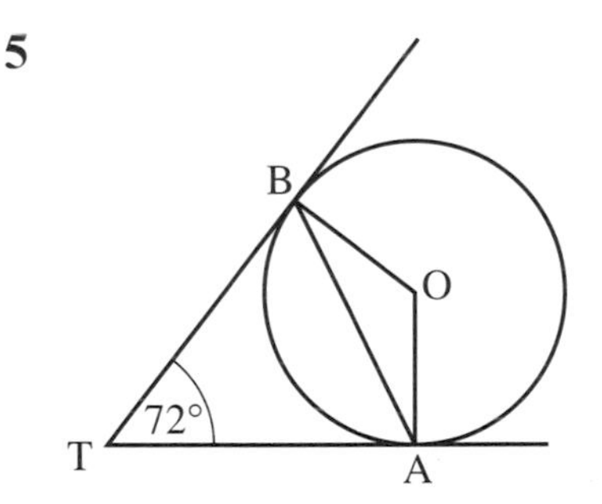

O is the centre of the circle
TA and TB are tangents to the circle
Angle ATB = 72°
Calculate the size of **(i)** ∠BAT
(ii) ∠OBA

6 Calculate the marked angles in these diagrams. O is always the centre of the circle. The lines drawn to the circles are tangents.

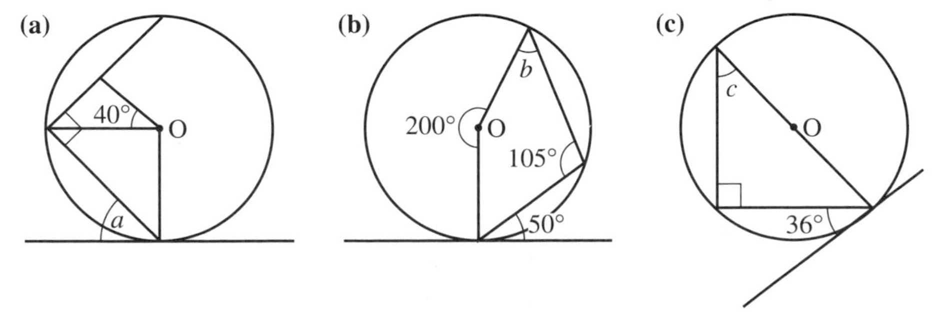

7 In all questions O is the centre of the circle. Calculate the named angles, giving reasons.

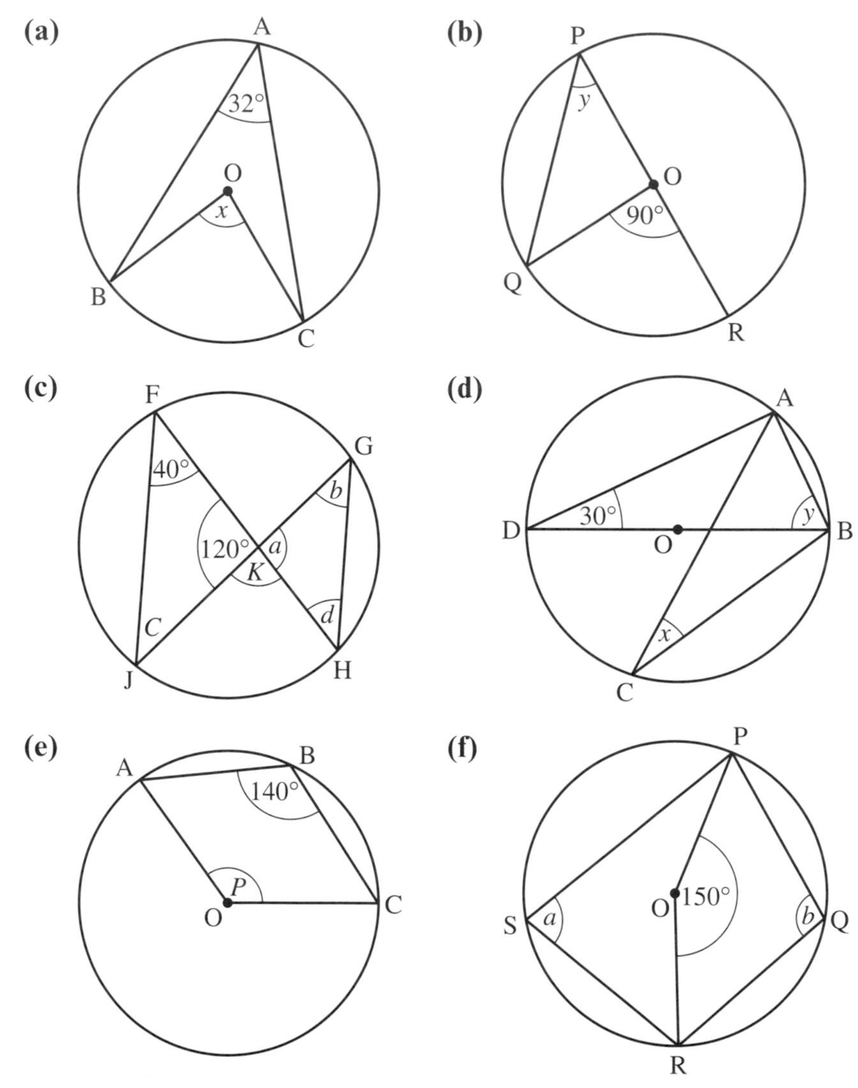

Exercise 10.3 Links: (*10E*) 10F

1 Write down the three-figure bearing for each of these directions.

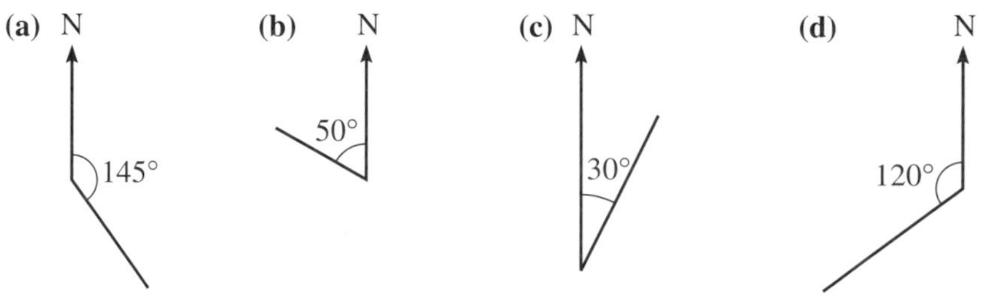

2 Write down the three-figure bearing for each of these directions:

(a) South **(b)** West **(c)** East **(d)** South West
(e) North East **(f)** North West **(g)** North **(h)** South East

3 Measure and write down the bearings of B from A in each of these diagrams.

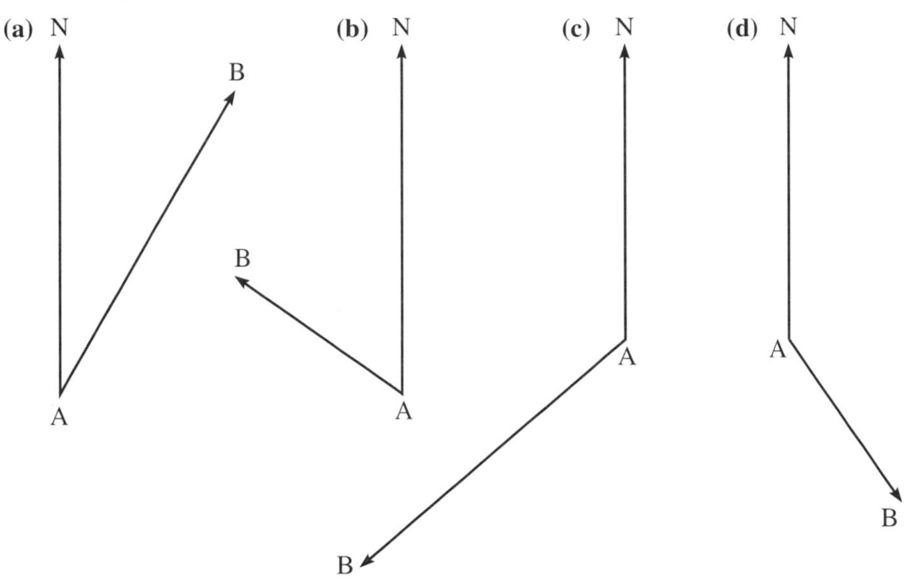

4 For each of the bearings in question **3** work out the bearing of A from B.

5 Draw diagrams to show the following bearings:
(a) Y on a bearing of 060° from X
(b) P on a bearing of 145° from Q
(c) R on a bearing of 310° from S
(d) G on a bearing of 220° from H

Exercise 10.4 Links: (*10F*) 10G

1 Calculate the interior angles of regular shapes with
(i) 8 sides **(ii)** 12 sides **(iii)** 9 sides **(iv)** 5 sides

2 Work out the sizes of the lettered angles in these diagrams.

c 54°
b 54° *a*
d

3 Make a copy of this shape.

Work out the sizes of all the angles in this shape and write them on your diagram.

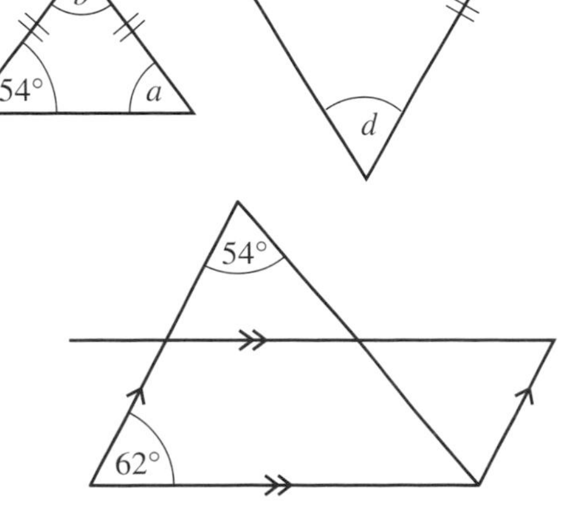

4 O is the centre of these circles.
Work out the size of the lettered angles.

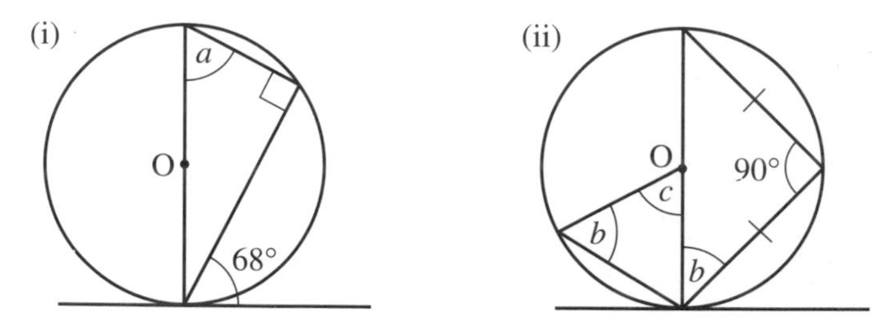

5 A ship travels on a bearing of 210°. It then turns around and travels back along the same path. Work out the new bearing.

6 A is 8 km due north of B. A ship leaves A and travels on a bearing of 120°. Another ship leaves B and travels on a bearing of 068°. Using a scale of 1 cm to represent 1 km draw a scale drawing and use it to find how far from A the ships paths cross.

11 Fractions

Exercise 11.1 Links: (*11A–C*) 11A–C

1 For each of these diagrams write down at least two equivalent fractions that describe the fraction shaded.

(a)

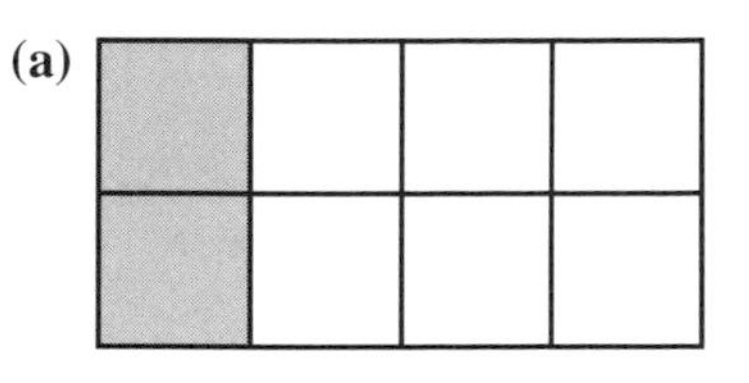

(b)

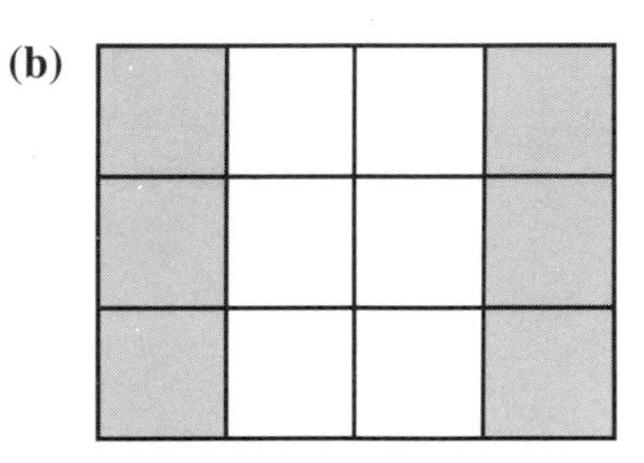

(c)

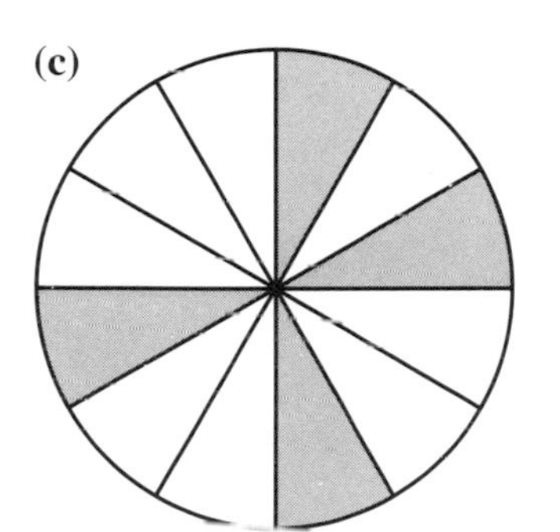

(d)

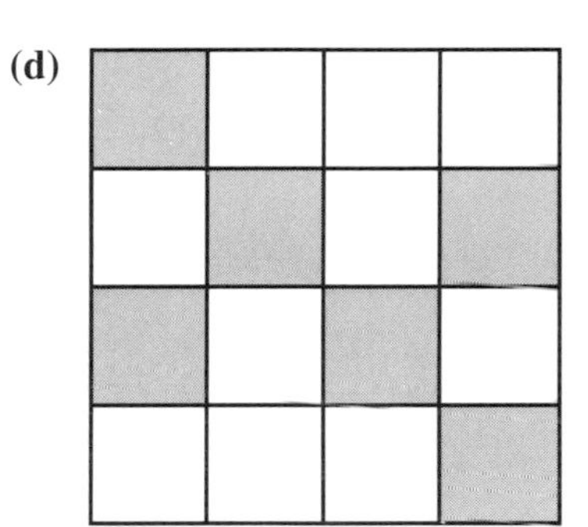

(e)

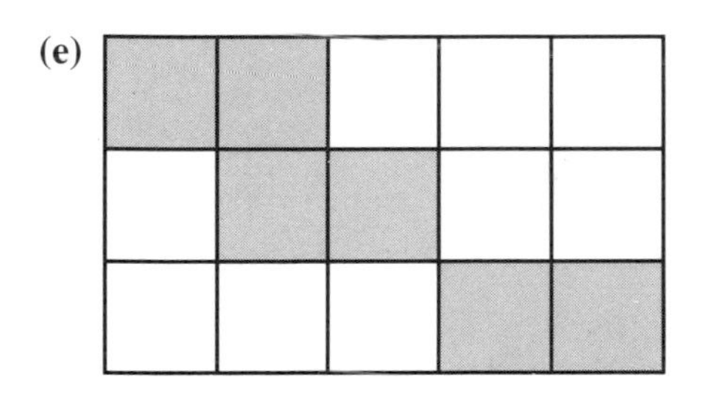

(f)

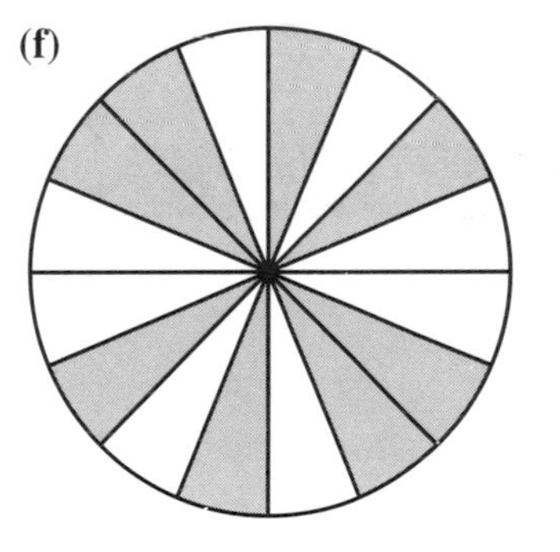

2 Copy and complete these sets of equivalent fractions:

(a) $\frac{1}{4} = \frac{}{8} = \frac{}{12} = \frac{}{16} = \frac{}{20} = \frac{}{24} = \frac{}{48}$

(b) $\frac{5}{6} = \frac{}{12} = \frac{}{18} = \frac{}{24} = \frac{}{30} = \frac{}{36} = \frac{}{72}$

(c) $\frac{3}{4} = \frac{}{100}$ **(d)** $\frac{2}{9} = \frac{6}{}$ **(e)** $\frac{5}{7} = \frac{}{35}$

(f) $\frac{7}{9} = \frac{49}{}$ **(g)** $\frac{4}{5} = \frac{}{45}$ **(h)** $\frac{3}{11} = \frac{}{55}$

(i) $\frac{5}{12} = \frac{30}{}$ **(j)** $\frac{2}{7} = \frac{}{21}$ **(k)** $\frac{9}{13} = \frac{}{39}$

(l) $\frac{1}{10} = \frac{7}{}$ **(m)** $\frac{4}{11} = \frac{16}{}$ **(n)** $\frac{6}{8} = \frac{}{4}$

3 Write each fraction in its simplest form:

(a) $\frac{6}{9}$ **(b)** $\frac{15}{20}$ **(c)** $\frac{18}{27}$ **(d)** $\frac{21}{28}$ **(e)** $\frac{4}{24}$
(f) $\frac{8}{36}$ **(g)** $\frac{16}{48}$ **(h)** $\frac{9}{45}$ **(i)** $\frac{32}{60}$ **(j)** $\frac{12}{90}$
(k) $\frac{25}{60}$ **(l)** $\frac{32}{48}$ **(m)** $\frac{19}{76}$

Remember:
'simplest form' means 'lowest terms'

Exercise 11.2 Links: (*11D, E*) 11D, E

Change to improper fractions:

1 **(a)** $1\frac{3}{4}$ **(b)** $2\frac{1}{5}$ **(c)** $4\frac{1}{8}$ **(d)** $3\frac{2}{3}$
(e) $5\frac{4}{5}$ **(f)** $3\frac{1}{2}$ **(g)** $6\frac{7}{10}$ **(h)** $2\frac{7}{12}$

2 **(a)** $4\frac{7}{8}$ **(b)** $7\frac{4}{9}$ **(c)** $3\frac{11}{12}$ **(d)** $4\frac{3}{8}$
(e) $2\frac{1}{12}$ **(f)** $5\frac{3}{10}$ **(g)** $11\frac{1}{2}$ **(h)** $6\frac{2}{3}$

Write each of the following fractions in its simplest form by changing it to a mixed number:

3 **(a)** $\frac{7}{5}$ **(b)** $\frac{9}{7}$ **(c)** $\frac{19}{5}$ **(d)** $\frac{21}{8}$
(e) $\frac{11}{5}$ **(f)** $\frac{13}{4}$ **(g)** $\frac{31}{8}$ **(h)** $\frac{14}{5}$

4 **(a)** $\frac{32}{7}$ **(b)** $\frac{15}{4}$ **(c)** $\frac{49}{8}$ **(d)** $\frac{83}{9}$
(e) $\frac{67}{12}$ **(f)** $\frac{41}{10}$ **(g)** $\frac{59}{8}$ **(h)** $\frac{43}{5}$

Exercise 11.3 Links: (*11F, G, H*) 11F, G, H

Write these fractions as decimals:

1 **(a)** $\frac{3}{5}$ **(b)** $\frac{7}{8}$ **(c)** $\frac{6}{10}$ **(d)** $\frac{13}{5}$
(e) $1\frac{1}{3}$ **(f)** $2\frac{3}{5}$ **(g)** $\frac{19}{25}$ **(h)** $\frac{8}{25}$

2 **(a)** $\frac{23}{1000}$ **(b)** $5\frac{3}{8}$ **(c)** $4\frac{17}{20}$ **(d)** $9\frac{16}{25}$
(e) $5\frac{13}{20}$ **(f)** $6\frac{7}{8}$ **(g)** $9\frac{19}{20}$ **(h)** $2\frac{1}{8}$

Write these decimals as fractions in their simplest form:

3 **(a)** 0.75 **(b)** 0.56 **(c)** 1.9 **(d)** 3.25
(e) 4.1875 **(f)** 3.45 **(g)** 10.004 **(h)** 5.625

4 **(a)** 0.001 25 **(b)** 4.504 **(c)** 14.75 **(d)** 15.15
(e) 7.5625 **(f)** 2.4375 **(g)** 8.075 **(h)** 6.969

5 Write these fractions as recurring decimals. Write your answers:
(i) as shown on the calculator display
(ii) using recurring decimal notation.

(a) $\frac{2}{3}$ **(b)** $1\frac{7}{9}$ **(c)** $4\frac{3}{11}$ **(d)** $2\frac{5}{6}$
(e) $3\frac{2}{9}$ **(f)** $\frac{56}{63}$ **(g)** $\frac{7}{13}$ **(h)** $4\frac{13}{44}$

Exercise 11.4 Links: (*11I, J*) 11I, J

1 Work out:

(a) $\frac{1}{4}+\frac{3}{4}$ **(b)** $\frac{3}{5}+\frac{4}{5}$ **(c)** $\frac{3}{10}+\frac{9}{10}$
(d) $3\frac{2}{5}+\frac{4}{5}$ **(e)** $2\frac{7}{8}+\frac{5}{8}$ **(f)** $\frac{4}{5}+\frac{4}{5}+\frac{1}{5}$
(g) $\frac{7}{10}+\frac{7}{10}+\frac{3}{10}$ **(h)** $2\frac{1}{2}+3\frac{3}{4}$ **(i)** $2\frac{7}{8}+3\frac{1}{2}$
(j) $4\frac{1}{5}+3\frac{1}{4}$ **(k)** $3\frac{1}{3}+4\frac{2}{5}$ **(l)** $2\frac{5}{6}+3\frac{2}{3}$
(m) $4\frac{5}{9}+2\frac{1}{3}$ **(n)** $3\frac{2}{5}+1\frac{4}{9}$ **(o)** $1\frac{11}{12}+2\frac{3}{4}$

2 Work out the perimeter of this rectangle.

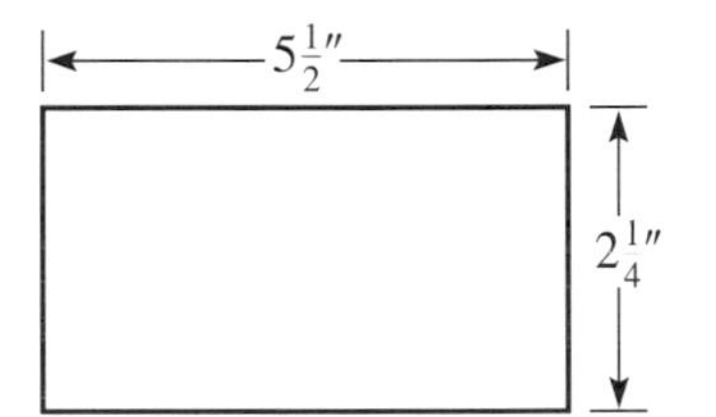

3 A bag which contains sweets weighs $2\frac{2}{5}$ lb. The sweets weigh $2\frac{1}{4}$ lb. What does the bag weigh?

4 Work out:

(a) $\frac{4}{5}-\frac{1}{5}$ **(b)** $\frac{7}{10}-\frac{2}{10}$ **(c)** $\frac{11}{12}+\frac{7}{12}$
(d) $\frac{3}{4}-\frac{3}{8}$ **(e)** $\frac{7}{8}-\frac{1}{4}$ **(f)** $\frac{11}{12}-\frac{1}{2}$
(g) $2\frac{1}{4}-1\frac{5}{8}$ **(h)** $3\frac{1}{3}-1\frac{5}{6}$ **(i)** $3\frac{7}{8}-1\frac{5}{6}$
(j) $4\frac{3}{8}-2\frac{1}{2}$ **(k)** $5\frac{1}{5}-3\frac{2}{3}$ **(l)** $4\frac{5}{9}-3\frac{5}{6}$
(m) $5\frac{1}{3}-1\frac{7}{9}$ **(n)** $8\frac{5}{6}-\frac{17}{18}$ **(o)** $6\frac{6}{7}-3\frac{4}{9}$

5 Bill spends $\frac{1}{3}$ of the day working and $\frac{1}{4}$ of the day sleeping. What fraction of the day is left?

6 A car and its passengers weigh $652\frac{1}{2}$ kg. The passengers weigh $45\frac{1}{2}$ kg, $48\frac{3}{4}$ kg, $56\frac{1}{4}$ kg and 72 kg. What is the weight of the car?

Exercise 11.5 Links: (*11K, L*) 11K, L

1 Work out:

(a) $\frac{3}{4} \times \frac{1}{2}$ **(b)** $\frac{2}{5} \times \frac{2}{3}$ **(c)** $\frac{5}{6} \times \frac{3}{5}$

(d) $\frac{7}{8} \times \frac{2}{3}$ **(e)** $\frac{1}{10} \times \frac{1}{10}$ **(f)** $\frac{11}{12} \times \frac{4}{9}$

(g) $\frac{4}{5} \times \frac{10}{11}$ **(h)** $\frac{9}{10} \times \frac{11}{15}$ **(i)** $1\frac{3}{4} \times 2\frac{1}{2}$

(j) $1\frac{1}{2} \times 2\frac{7}{8}$ **(k)** $2\frac{3}{5} \times 1\frac{2}{3}$ **(l)** $5\frac{6}{7} \times 2\frac{4}{5}$

2 A car wash takes $\frac{1}{10}$ hour to wash one car. How long does it take to wash 26 cars?

3 A round of golf takes $3\frac{1}{2}$ hours to play. How long will 6 rounds take?

4 A book weighs $2\frac{1}{4}$ lb. How much do 8 books weigh?

5 Work out:

(a) $\frac{1}{4}$ of 36 **(b)** $\frac{7}{8}$ of 40 **(c)** $\frac{5}{6}$ of £6.72

(d) $\frac{5}{7}$ of 630 sweets **(e)** $\frac{2}{7}$ of £17.01 **(f)** $\frac{5}{9}$ of 1557

6 Lucy earns £3.35 per hour. She works for 12 hours and she spends $\frac{3}{5}$ of this. How much does she spend?

7 Les earns £156.40 in one week. He pays $\frac{1}{5}$ of this amount in tax. How much tax does he pay?

Exercise 11.6 Links: (*11M, N*) 11M, N

1 Work out:

(a) $\frac{4}{9} \div \frac{2}{3}$ **(b)** $\frac{3}{4} \div \frac{5}{9}$ **(c)** $\frac{5}{8} \div \frac{2}{3}$

(d) $\frac{11}{12} \div 3$ **(e)** $\frac{5}{6} \div 4$ **(f)** $3\frac{3}{5} \div 6$

(g) $5\frac{1}{5} \div 13$ **(h)** $2\frac{1}{3} \div 1\frac{1}{6}$ **(i)** $1\frac{1}{5} \div 3\frac{2}{3}$

(j) $2\frac{2}{3} \div 1\frac{3}{5}$ **(k)** $1\frac{1}{3} \div 7\frac{1}{5}$ **(l)** $3\frac{7}{8} \div 2\frac{5}{12}$

2 A salt cellar holds $2\frac{1}{2}$ ounces of salt. A tub of salt holds 20 ounces. How many salt cellars can be filled from the tub?

3 Find the difference between $\frac{4}{5}$ of 45 miles and $\frac{7}{8}$ of 48 miles.

4 On a coach 25 of the passengers are women, 15 are men and 5 are children. What fraction of the total number of people are women?

5 A metal bar is 120 cm long. When heated it expands by $\frac{3}{8}$ of its original length. What is its length when heated?

6 Rodger had 96 sheep last year. His flock increased by $\frac{3}{8}$ this year.
How many sheep does Rodger have now?

7 A sack of potatoes weighs 56 kg. More potatoes are added and the weight of the sack of potatoes increases by $\frac{3}{7}$.
What is the new weight of the sack of potatoes?

Exercise 11.7 Links: (*11O*) 11O

1 Copy and complete these sets of equivalent fractions:

(a) $\frac{5}{8} = \frac{}{16} = \frac{}{24} = \frac{}{32} = \frac{}{40} = \frac{}{48} = \frac{}{64}$

(b) $\frac{2}{7} = \frac{}{14} = \frac{}{21} = \frac{}{28} = \frac{}{35} = \frac{}{63} = \frac{}{84}$

2 Express these fractions in their simplest form:

(a) $\frac{24}{36}$ **(b)** $\frac{23}{8}$ **(c)** $\frac{63}{8}$

3 Express these fractions as decimals:

(a) $\frac{5}{9}$ **(b)** $3\frac{7}{8}$ **(c)** $2\frac{5}{11}$ **(d)** $5\frac{19}{20}$

4 Express these decimals as fractions:

(a) 0.15 **(b)** 2.625 **(c)** 0.66 **(d)** 0.0072

5 Find the difference between $\frac{3}{4}$ of 32 and $\frac{2}{3}$ of 27.

6 Henlow to Hitchin is $6\frac{1}{2}$ miles. Clifton to Hitchin is $7\frac{1}{3}$ miles.
How much further is it to Hitchin from Clifton than from Henlow?

7 There are 56 kg of potatoes in a sack. How many $2\frac{1}{2}$ kg bags can be filled from one sack?

8 A television was reduced from £320 to £280 in a sale. By what fraction was the price reduced in the sale?

9 Helen spent £12 on make up, £16 on shoes and £32 on clothes.
What fraction of the total amount did she spend on

(a) make up
(b) shoes
(c) clothes?

10 The second hole on a golf course is 480 yards.
A golfer hits the golf ball 220 yards towards the hole.
What fraction of the total distance did he hit the ball?

12 Measure 2

Exercise 12.1 — Links: (*12A – D*) 12A – D

1 Change these weights into grams:
(a) 3 kg **(b)** 2.1 kg **(c)** 5.81 kg
(d) 0.012 kg **(e)** 10.0189 kg

2 Change these weights into kilograms:
(a) 10 000 g **(b)** 3 tonnes **(c)** 4800 g
(d) 500 g **(e)** 60 g

3 Change these lengths into centimetres:
(a) 3.5 m **(b)** 160 mm **(c)** 2.83 m
(d) 25 mm **(e)** 1 km

4 Change these lengths into millimetres:
(a) 3 cm **(b)** 4.1 cm **(c)** 0.55 cm
(d) 1.24 m **(e)** 6.05 cm

5 Change these lengths into metres:
(a) 3.7 km **(b)** 870 cm **(c)** 2650 mm
(d) 1.937 km **(e)** 1005 cm

6 Put these weights into order of size with the smallest first:
420 g, 4 kg, 39.5 kg, 4220 mg, 0.405 kg

7 Put these lengths into decreasing order of size:
1 km, 900 m, 1200 m, 11 000 cm, 1 050 000 mm

Exercise 12.2 — Links: (*12E, F*) 12E, F

1 Mary usually fills up with 7 gallons of petrol. The pump is calibrated in litres.
How many litres should Mary buy?

2 Given that 1 inch = 2.54 cm, 12 inches = 1 foot, 3 feet = 1 yard, change 4 yards into metres.

3 In this question take 14 lbs = 1 stone and 2.2 lbs = 1 kg.
Karl weighs 11 stone 10 lbs. Work out his weight in kg.

4 25 g = 1 ounce and there are 16 ounces in 1 lb. A bag of potatoes weighs 5 kg. Work out what this is in lbs and ounces.
Work out the weight in lbs if the conversion used is 1 kg = 2.2 lbs.

5 Use the information 30 cm = 1 foot, 1 mile = 5280 feet, to work out how many metres there are in a mile.

6 The distance from London to Aberdeen by rail is 530 miles. How far is this in kilometres?

7 A plastic beaker holds $\frac{1}{3}$ pint. How many of these beakers can be filled from a 5-litre container? Use 1 litre = 1.75 pints.

8 Which is longer and by how much: 15 inches or 38 cm?

9 Stephen has a waist measurement of 34 inches. All the measurements on clothes at a shop are in centimetres. What value of cm should he ask for?

10 1 hundredweight (112 lbs) of rice is weighed into 200 g packets. How many can be filled?

11 Gary is 1.86 metres tall. How tall is he in feet and inches? Use 12 inches = 1 foot and 2.54 cm = 1 inch.

12 A chain is an old English measurement of length. It is 22 yards long. How far is this in metres? Give your answer to 1 decimal place.

13 What is the difference in weight between an Imperial ton and a metric tonne? There are 2240 lbs in an Imperial ton. Give your answer **(a)** in pounds (lbs) and **(b)** in kg.

Exercise 12.3 Links: (*12G, H*) 12G, H

1 The area of a flat roof is 60 m^2. How many litres of water fall on the roof during a downpour of 1 mm? (Hint: work in cm.)

2 Work out the volume of a cuboid which measures 2 m by 50 cm by 50 cm.
Give your answer in **(a)** cm^3 and **(b)** m^3.

3 A school field is 6 hectares in area. The annual rainfall for the region is 20 inches.
Find **(a)** in cubic metres **(b)** in gallons the volume of water that falls on the field in 1 year. Use 1 hectare = 100 m by 100 m, 1 inch = 2.54 cm, 1 m^3 = 220 gallons.

4 A rectangular field measures 73 yards by 124.2 yards. Given that 1 yard = 0.914 metres, work out the metric area of the field.
Give your answers in **(a)** m^2 and **(b)** hectares to 2 d.p.

5 There are 21 shillings in a guinea.
There are 5 shillings in a crown.
There are 3 groats in a shilling.
(a) How many shillings are there in 5 guineas?
(b) How many groats are there in 2 guineas?
(c) How many groats are there in 3 crowns?
(d) How many shillings are there in 4 crowns?
(e) How many crowns are there in 15 guineas?

6 Using the information given in question **5**, for each of the following work out which is worth most and by how much.
(a) 8 crowns or 2 guineas
(b) 5 crowns and 1 groat or 1 guinea and 5 shillings
(c) 12 shillings and 2 crowns or 1 guinea and 3 groats

Exercise 12.4 Links: (*12I, J*) 12I, J

1 These weights are given to the nearest gram.
Write down the greatest and least weight they could be.
(a) 153 g **(b)** 210 g **(c)** 3.000 kg **(d)** 16 g **(e)** 400 g

2 These race times are correct to the nearest $\frac{1}{5}$ second.
Write down the greatest and least times they could be.
(a) 10.2 secs **(b)** 21.6 secs **(c)** 41.4 secs **(d)** 50.0 secs

3 State suitable units for measuring:
(a) the volume of a bucket
(b) the height of a mountain
(c) the weight of cucumber
(d) the time to boil an egg
(e) the area of a field
(f) the area of Scotland
(g) the weight of an ocean liner.

Exercise 12.5 Links: (*12K, L, M*) 12K, L, M

1 Rose walked 1400 metres in 20 minutes.
What was her average speed?

2 The average speed of a train travelling from London to York is 150 km/h. The distance is 328 km.
How long does the journey take?

3 Copy and complete the table:

Journey	Distance in km	Journey time in hours/min	Average speed in km/h
London–Glasgow	642	5 h 22 min	120
Reading–Exeter	219	2 h 03 min	
Birmingham–Liverpool	141		85
York–Newcastle	128	59 min	
Cardiff–Southampton		2 h 29 min	74
Colchester–Norwich	101		93
Plymouth–Penzance	128	1 h 58 min	
Portsmouth–Brighton		1 h 33 min	46
Leeds–Carlisle	181		68
Bristol–Swansea	130	1 h 27 min	
London–York		1 h 51 min	163

4 Copy and complete the table:

Country	Area in km^2	Population (1991)	Population density
Nepal	147 000	18.5 million	126
Norway	324 000	4.25 million	
Paraguay	407 000	million	10.1
Romania	229 000	22.8 million	
Sierra Leone	71 700	3.52 million	
Sudan	2 510 000	million	8.2
Sweden		8.59 million	20.9
Syria	184 000	9.05 million	
U.K.		56.5 million	233.3
Barbados	431	259 thousand	
Jamaica	11 000	million	225.4

5 The density of the Earth is $5.5\,g/cm^3$. The Earth is almost a sphere with radius 6378 km.
Calculate an estimate for the mass of the Earth.
Volume of a sphere $= \frac{4}{3} \times \pi r^3$.

Exercise 12.6 Links: (*12N, O*) 12N, O

1 Change 40 metres per second to kilometres per hour.

2 Change 25 kilometres per hour to metres per second.

3 Change 1800 miles per hour to miles per second.

4 Change 50 centimetres per second to metres per minute.

5 The speed of sound at sea level is 760 miles per hour.
(a) Change this into kilometres per hour.
(b) What is the speed of sound at sea level in metres per second.
(c) Is 20 kilometres per minute faster than the speed of sound?

6 The density of beech wood is 700 kg per m^3.
The volume of a beech wood statue is 720 cm^3.
Work out the weight of the statue.

7 The density of horse chestnut wood is 540 kg per m^3.
(a) Change this to g per cm^3.
(b) How much does a cuboid of horse chestnut wood measuring 5 cm by 15 cm by 20 cm weigh?

Exercise 12.7 Links: (*12P*) 12P

1 A person on a business trip travels from London to Birmingham (184 km), Birmingham to Nottingham (68 miles) and from Nottingham back to London (126 miles).
Work out the total distance travelled in **(a)** miles
and **(b)** kilometres

2 A small gold bar is in the shape of a cuboid. The cuboid is 6 cm wide, 15 cm long and 4 cm high.
Given that 1 cm^3 of gold weighs 19.35 g, work out the weight of the gold bar in kilograms.

3 28.4 grams = 1 ounce   16 ounces = 1 pound
14 pounds = 1 stone

(a) Find how many grams are equivalent to 1 stone.
(b) Work out, in pounds and ounces to the nearest ounce, the weight of a baby weighing 3720 g.

4 Roger Bannister was the first man to run a mile in 4 minutes.
Work out his average speed in km h^{-1} and metres per second.

13 Quadratic functions

Exercise 13.1 **Links: (*13A*, *B*) 13A, B**

1 Write down the functions for each of these number machines:

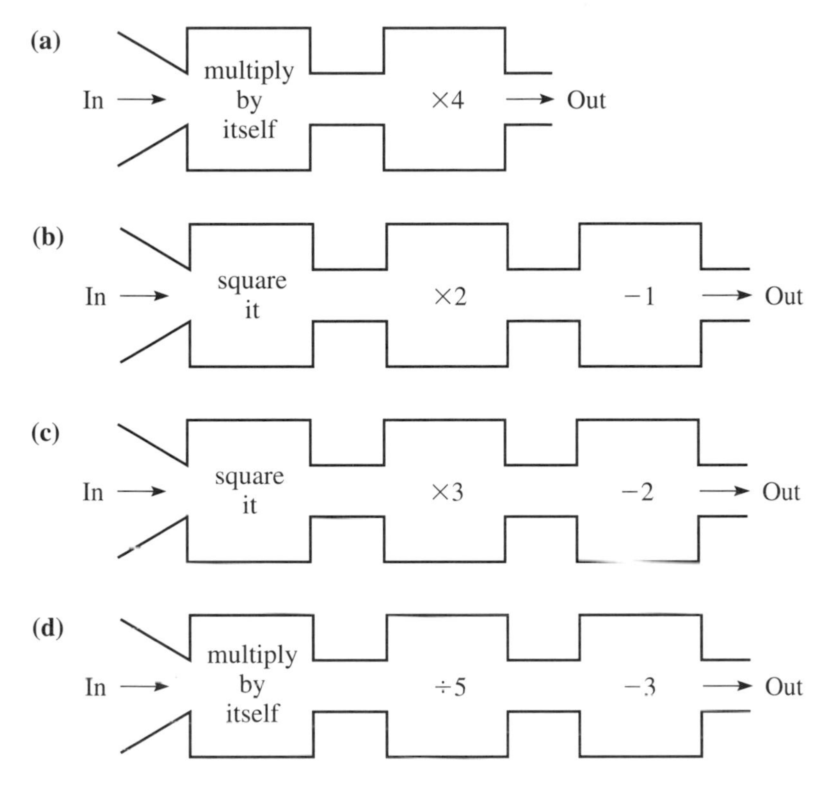

2 The inputs to each of these functions are:

$n = 0, 1, 2, 3, 4, 5$

Work out the outputs in each case.

(a) $n \rightarrow n^2 + 3$ **(b)** $n \rightarrow 2n^2$ **(c)** $n \rightarrow 2n^2 + 3$

(d) $n \rightarrow 2n^2 - 3$ **(e)** $n \rightarrow 3n^2 + \frac{1}{2}$ **(f)** $n \rightarrow \frac{1}{2}n^2 + 5$

3 Put the values 0, 1, 2, 3, 4, 5 into each of these quadratic functions.

(a) $n \rightarrow n^2 + 3$ **(b)** $n \rightarrow 3n^2$ **(c)** $n \rightarrow 3n^2 + 1$

(d) $n \rightarrow 4n^2 - 1$ **(e)** $n \rightarrow \frac{1}{2}n^2 + 2$ **(f)** $n \rightarrow -3n^2 + 1$

Draw a mapping diagram for each set of results.

Work out the first and second differences in each case. In each case comment on:

(i) the output when the input is 0

(ii) the second differences.

4 A quadratic function is of the type:

$n \rightarrow 3n^2 + c$

When the input is 0 the output is 2.

(a) Work out the value of c.

(b) What would be the second differences if the inputs were 0, 1, 2, 3, 4, 5?

5 A quadratic function is of the type:

$n \rightarrow an^2 + c$

(a) When the input is $n = 0$, the output is 5. Find the value of c.

(b) When the input is $n = 1$, the output is 7. Find the value of a.

(c) Work out the output when the input is:

(i) $n = 2$ **(ii)** $n = 3$ **(iii)** $n = 10$

6 A quadratic function is of the type:

$n \rightarrow an^2 + c$

When the input is 0 the output is 2.
When the input is 1 the output is 4.

(a) Work out the value of:

(i) c **(ii)** a

(b) Work out the output when the input is:

(i) 2 **(ii)** 3 **(iii)** 5 **(iv)** 12

7 A quadratic function is of the type:

$n \rightarrow an^2 + c$

When the input is 0 the output is -1.
When the input is 1 the output is 2.
Work out the output when the input is:

(i) 1 **(ii)** 5 **(iii)** 10

Exercise 13.2 Links: (*13C, D*) 13C, D

1 Put the values for n of $n = 0, 1, 2, 3, 4, 5$ into each of these larger quadratic functions.
In each case draw a mapping diagram and work out the first and second differences.

(a) $n \rightarrow 2n^2 + 11n + 5$ **(b)** $n \rightarrow 3n^2 + 2n - 4$

(c) $n \rightarrow \frac{1}{2}n^2 + 3n + 1$ **(d)** $n \rightarrow \frac{1}{4}n^2 + 3n + 5$

(e) $n \rightarrow -n^2 + n + 3$ **(f)** $n \rightarrow -3n^2 + 2n - 1$

In each case, form a difference table.
Use the difference table to confirm that for a quadratic function of the type:

$n \rightarrow an^2 + bn + c$

the second differences are equal to $2a$.

2 Each of these patterns of inputs → outputs has been produced by a quadratic function. Find the function for each pattern.

(a)	(b)	(c)	(d)
$0 \to 0$	$0 \to 5$	$0 \to -1$	$0 \to 0$
$1 \to 1$	$1 \to 6$	$1 \to 2$	$1 \to 2$
$2 \to 4$	$2 \to 9$	$2 \to 11$	$2 \to 6$
$3 \to 9$	$3 \to 14$	$3 \to 26$	$3 \to 12$
$4 \to 16$	$4 \to 21$	$4 \to 47$	$4 \to 20$
$5 \to 25$	$5 \to 30$	$5 \to 74$	$5 \to 30$

(e)	(f)	(g)	(h)
$0 \to 0$	$0 \to 0$	$0 \to 1$	$0 \to 1$
$1 \to 3$	$1 \to -2$	$1 \to 3$	$1 \to 6$
$2 \to 8$	$2 \to -2$	$2 \to 7$	$2 \to 15$
$3 \to 15$	$3 \to 0$	$3 \to 13$	$3 \to 28$
$4 \to 24$	$4 \to 4$	$4 \to 21$	$4 \to 45$
$5 \to 35$	$5 \to 10$	$5 \to 31$	$5 \to 66$

In each case, work out the output when the input is 10.

Exercise 13.3 Links: (*13E*) 13E

1 The first five terms of some sequences are given below. In each case work out, in terms of n, an expression for the nth term of the sequence:

(a) 3, 6, 11, 18, 27, ...
(b) -2, 1, 6, 13, 22, ...
(c) 2, 8, 18, 32, 50, ...
(d) 4, 13, 28, 49, 76, ...
(e) $2\frac{1}{2}$, 4, $6\frac{1}{2}$, 10, $14\frac{1}{2}$, ...
(f) -1, 0, 3, 8, 15, ...
(g) 0, 3, 8, 15, 24, ...
(h) 4, 8, 14, 22, 32, ...
(i) 4, 11, 22, 37, 56, ...
(j) 12, 28, 50, 78, 112, ...

2 The first five terms in a sequence are:

99, 96, 91, 84, 75

Find, in terms of n an expression for the nth term of this sequence.

3 Janice builds a pattern out of circular mats. The first 4 stages in her pattern are shown below.

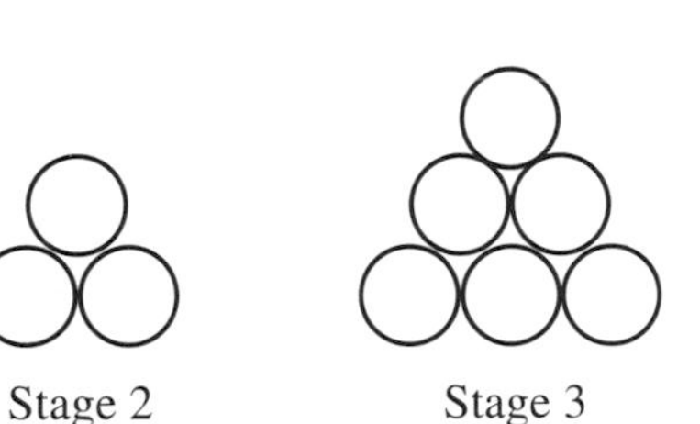

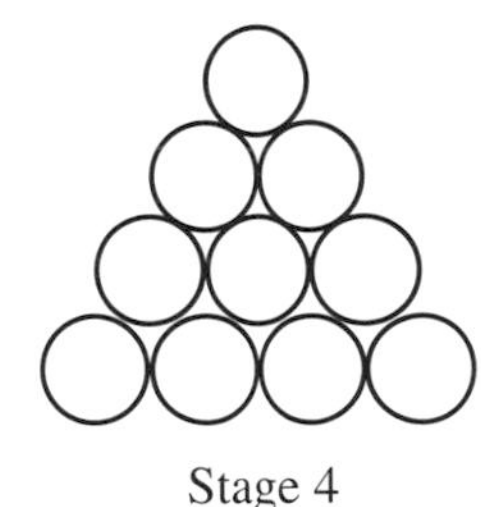

Stage 1 Stage 2 Stage 3 Stage 4

(a) Draw the pattern for Stage 5.
(b) Copy and complete the table below.

Stage	1	2	3	4	5	6	7	8	9	10
Number of mats	1	3	6	10						

(c) Find, in terms of n, an expression for the number of mats in Stage n.

4 Here are the first 4 stages in a pattern made from small squares

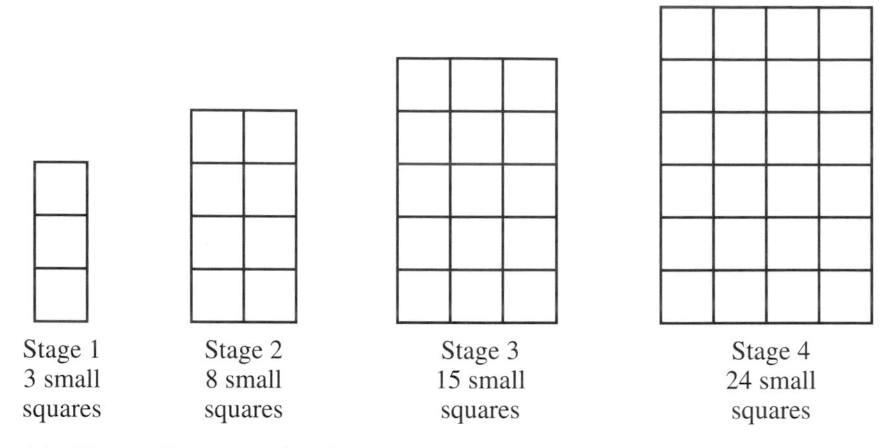

Stage 1 3 small squares | Stage 2 8 small squares | Stage 3 15 small squares | Stage 4 24 small squares

(a) Draw Stage 5 of this pattern.
(b) How many small squares will there be in:
(i) Stage 6 of the pattern **(ii)** Stage 9 of the pattern.
(c) Find, in terms of n, an expression for the number of small squares in Stage n of the pattern.

5 The even numbers, starting with 2, are set out in rows:

		Sum of the numbers in the row
1st row	2	2
2nd row	2, 4	$2 + 4 = 6$
3rd row	2, 4, 6	$2 + 4 + 6 = 12$
4th row	2, 4, 6, 8	$2 + 4 + 6 + 8 = 20$

(a) Write down:
(i) the numbers in the 5th row
(ii) the sum of the numbers in the 5th row.
(b) Copy and complete the table below.

Row	1	2	3	4	5	6	7	8	9
Sum of the numbers in the row	2	6	12	20					

(c) Work out, in terms of n, an expression for the sum of the numbers in the nth row. [E]

6 The expression $\frac{n(n+1)}{2}$ is the nth term of the sequence of triangular numbers:

1, 3, 6, 10

Write down an expression, in terms of n, for the nth term of the sequence:

10, 30, 60, 100 [E]

7 These diagrams represent the first three trapezium numbers (each diagram always starts with 2 dots in the top row).

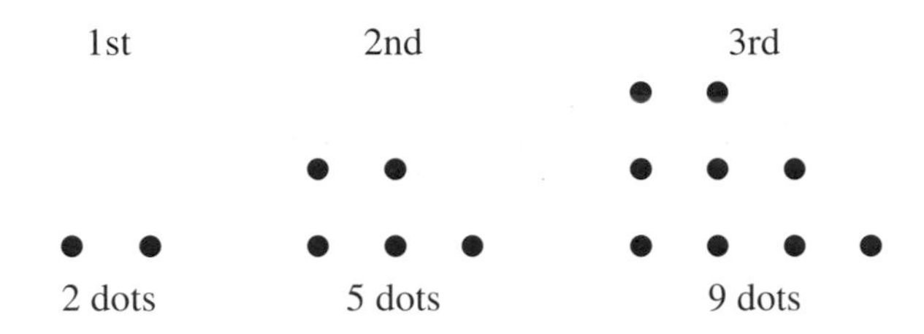

(a) Work out the number of dots in:
(i) the 4th **(ii)** the 5th **(iii)** the 6th trapezium number

(b) Find an expression, in terms of n, for the number of dots in the nth trapezium number. [E]

8 These diagrams show the first four stages of a pattern of dots.

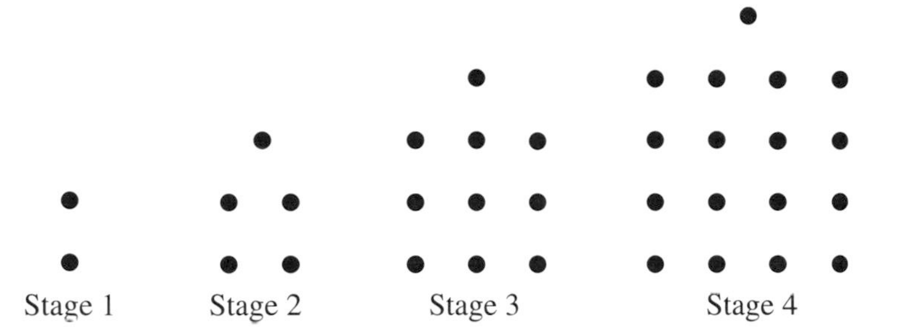

(a) Draw the pattern for Stage 5.

(b) Copy and complete the table.

Stage	1	2	3	4	5	6	7	8	9	10
Number of dots	2	5	10	17						

(c) Find, in terms of n, an expression for the number of dots in Stage n.

(d) Use your expression to work out
(i) the number of dots in Stage 50
(ii) the stage number when the number of dots is 145.

9 The first six terms in a sequence are $-2, -2, 0, 4, 10, 18$
Find an expression, in terms of n, for the nth term of this sequence.

14 Properties of numbers

Exercise 14.1 Links: (*14A, B*) 14A, B

1 Write down all the odd numbers between:
(a) 40 and 50 (b) 230 and 250

2 Write down all the even numbers between:
(a) 80 and 90 (b) 190 and 200

3 Write down all the multiples of 8 less than 100.

4 Write down all the factors of:
(a) 36 (b) 48

5 Write down all the prime numbers between:
(a) 5 and 20 (b) 60 and 80

6 Write down:
(a) the 5th square number
(b) the 3rd cube number
(c) the 18th square number
(d) the 9th cube number
(e) the 8th to 15th square numbers
(f) the 10th to 13th cube numbers.

7 From this list write down all the numbers which are:
(a) square numbers (b) cube numbers

100, 27, 1, 343, 625, 324, 96, 1331, 32, 125, 256

Exercise 14.2 Links: (*14C, D*) 14C, D

1 Use your calculator to work out:
(a) 14^2 (b) 4.6^2 (c) 36^3
(d) 17.2^2 (e) 0.012^3 (f) 13.9^2
(g) $(-4.3)^3$ (h) $\sqrt{361}$ (i) $\sqrt[3]{729}$
(j) $\sqrt[3]{1728}$ (k) $\sqrt{1296}$ (l) $\sqrt{0.0196}$

2 In this question give your answers correct to 3 significant figures.
(a) $\sqrt{198}$ (b) $\sqrt{3479}$ (c) $\sqrt{21.9}$
(d) $\sqrt{392\,761}$ (e) $\sqrt[3]{10}$ (f) $\sqrt[3]{-14.9}$
(g) $\sqrt[3]{47.8}$ (h) $\sqrt[3]{1219.7}$ (i) $\sqrt[3]{7\,846\,591}$

3 Use a trial and improvement method to find these roots correct to 2 decimal places. Use a calculator to check your answers.

(a) $\sqrt{6}$ **(b)** $\sqrt[3]{19}$ **(c)** $\sqrt{19}$
(d) $\sqrt[3]{32}$ **(e)** $\sqrt{60}$ **(f)** $\sqrt[3]{48}$
(g) $\sqrt{39}$ **(h)** $\sqrt[3]{40}$ **(i)** $\sqrt{27}$

Exercise 14.3 Links: (*14E, F*) 14E, F

Work out:

1 **(a)** 3^3 **(b)** 2^5 **(c)** 5^6

2 **(a)** 11^3 **(b)** 7^3 **(c)** 10^5

3 **(a)** 7^5 **(b)** $5^3 + 4^2$ **(c)** $12^3 \div 3^4$

Find the value of x:

4 **(a)** $2^x = 16$ **(b)** $4^x = 64$ **(c)** $5^x = 625$

5 **(a)** $2^x = 128$ **(b)** $7^x = 343$ **(c)** $4^x = 1024$

6 **(a)** $10^x = 10\,000\,000$ **(b)** $8^x = 4096$ **(c)** $6^x = 7776$

Simplify by writing as a single power of the number:

7 **(a)** $2^5 \times 2^3$ **(b)** $6^4 \times 6^6$ **(c)** $4^3 \times 4^7$

8 **(a)** $2^7 \div 2^4$ **(b)** $5^6 \div 5^1$ **(c)** $3^7 \div 3^6$

9 **(a)** $2^3 \times 2^4 \times 2^6$ **(b)** $7^2 \times 7^5 \times 7^4$ **(c)** $7^5 \div 7$

10 **(a)** $\dfrac{5^5 \times 5^4}{5^6}$ **(b)** $\dfrac{2^8 \times 2^5}{2^7}$ **(c)** $\dfrac{7^4 \times 7^6}{7^9}$

Exercise 14.4 Links: (*14G*) 14G

1 Write these numbers in prime factor form:

(a) 12 **(b)** 36 **(c)** 54 **(d)** 17

2 Find the highest common factor of:

(a) 4 and 12 **(b)** 16 and 24 **(c)** 14 and 70
(d) 42 and 28 **(e)** 72 and 30

3 Find the lowest common multiple of:

(a) 2 and 3 **(b)** 3 and 5 **(c)** 8 and 12
(d) 12 and 18 **(e)** 48 and 60

Exercise 14.5 Links: (*14H – K*) 14H – K

1 Write in standard form:
(a) 5100 **(b)** 700 000 **(c)** 4700
(d) 496 000 **(e)** 9 **(f)** 63 000 000

2 Write as ordinary numbers:
(a) 6×10^4 **(b)** 2.17×10^3 **(c)** 9.1×10^6
(d) 17.39×10^2 **(e)** 5.01×10^5 **(f)** 2.06×10^7

3 Write in standard form:
(a) 0.52 **(b)** 0.0076 **(c)** 0.0972
(d) 0.000 012 7 **(e)** 0.000 002 3 **(f)** 0.000 452

4 Write as ordinary numbers:
(a) 7×10^{-2} **(b)** 6.3×10^{-4} **(c)** 3.5×10^{-6}
(d) 8.017×10^{-4} **(e)** 4.92×10^{-3} **(f)** 9.06×10^{-7}

5 Work out these. Give your answers in standard form correct to 3 s.f.
(a) $(5.9 \times 10^3) + (9.6 \times 10^2)$ **(b)** $(8.17 \times 10^3) - (6.24 \times 10^2)$
(c) $(4.36 \times 10^5) \times (1.4 \times 10^{-3})$ **(d)** $(3.7 \times 10^{-5}) \times (9.1 \times 10^{-3})$
(e) $\dfrac{6.5 \times 10^9}{8.3 \times 10^8}$ **(f)** $\dfrac{3.5 \times 10^{12}}{9.7 \times 10^9}$ **(g)** $\dfrac{5.4 \times 10^3}{7.2 \times 10^5}$

6 A nanometre is 10^{-9} metres. Write 150 nanometres in metres. Give your answer in standard form.

7 Write down two prime numbers between 30 and 40.

8 Simplify **(a)** $5^4 \times 5^3$ **(b)** $5^6 \div 5^4$ **(c)** $\dfrac{5^4 \times 5^5}{5^6}$

9 Work out the highest common factor of:
(a) 15 and 25 **(b)** 42 and 48

10 Work out the lowest common multiple of:
(a) 5 and 7 **(b)** 12 and 20

11 The mass of a neutron is 1.675×10^{-24} grams. Calculate the mass of 250 000 neutrons. Give your answer in standard form.

12 Light travels at a speed of $3 \times 10^8\ \text{ms}^{-1}$. Work out how far, in metres, light will travel in 500 years. (Take one year as $365\frac{1}{4}$ days.) Give your answer in standard form.

15 Pythagoras' theorem

Exercise 15.1 — Links: (*15A*, *B*) 15A, B

1 Find the hypotenuse in these right-angled triangles:

(a) 4 cm, 3 cm

(b) 8 cm, 6 cm

(c) 12 cm, 9 cm

(d) 7 cm, 6 cm

(e) 8 cm, 7 cm

(f) 9 cm, 8 cm

2 Calculate the lengths marked with letters in these triangles:

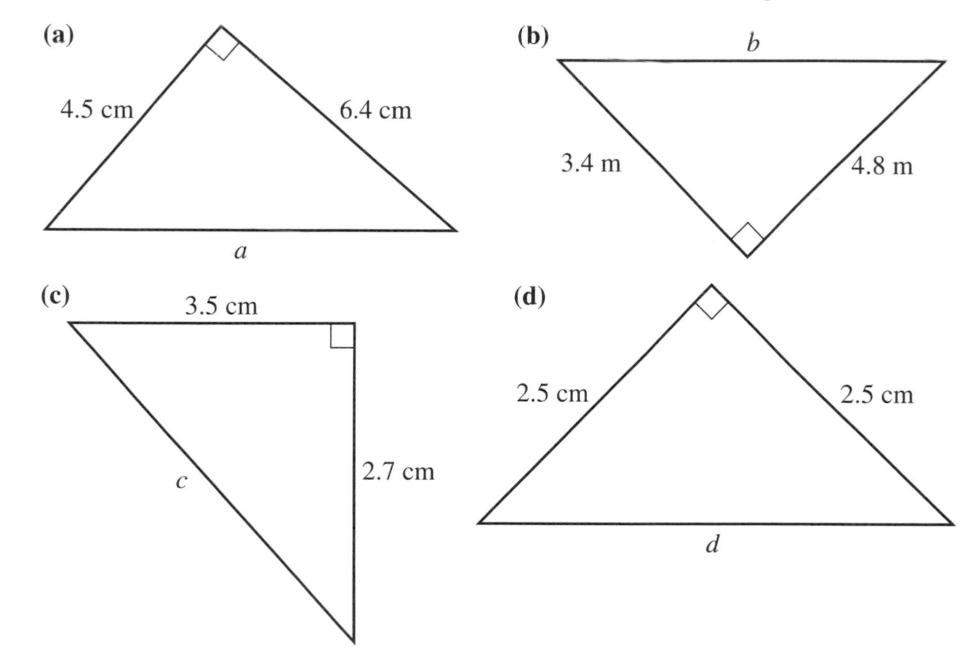

3 A ladder is resting against the wall of a house. The foot of the ladder is 3 m from the base of the wall and the top of the ladder is 4 m from the base of the wall. How long is the ladder?

4 Calculate the span of the roof truss shown in the diagram.

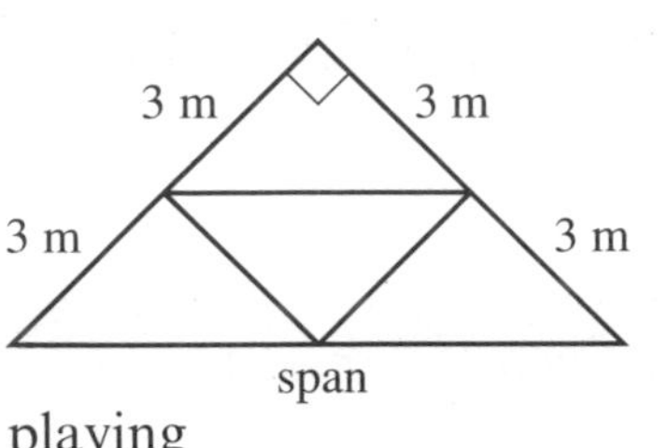

5 Susie is flying her kite on a horizontal playing field. The string is taut and the kite is 100 m above the ground. The kite is 300 m from Susie in a horizontal direction.
How long is the kite string?

100 m
300 m

Exercise 15.2 Links: (*15C*) 15C

1 Calculate the unmarked sides in these right-angled triangles:

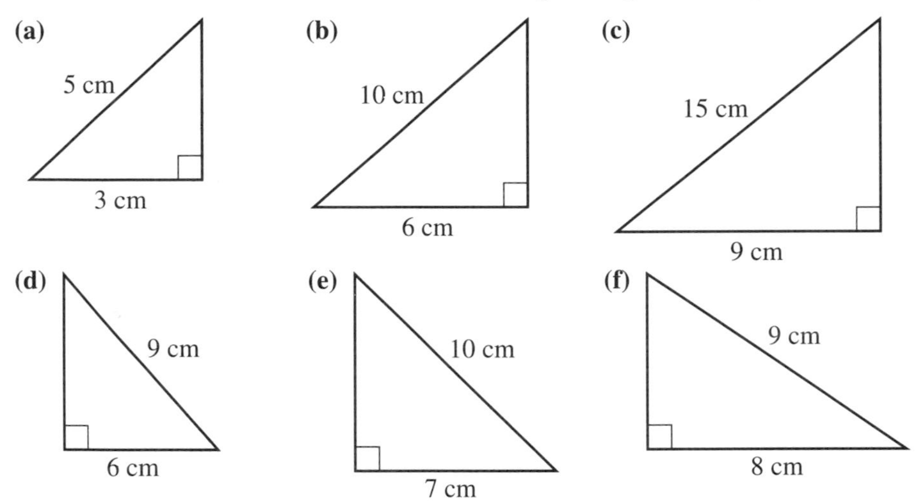

2 Calculate the lengths marked with letters in these triangles:

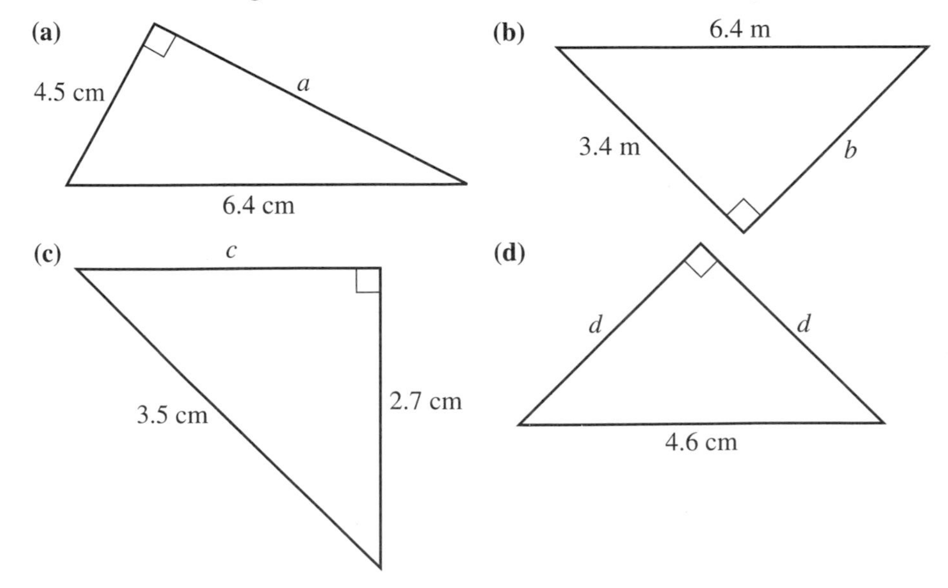

3 Keith used his 6-metre long ladder to clean his upstairs windows. He placed the ladder 2 metres away from the foot of the wall. How far up the wall did the ladder reach?

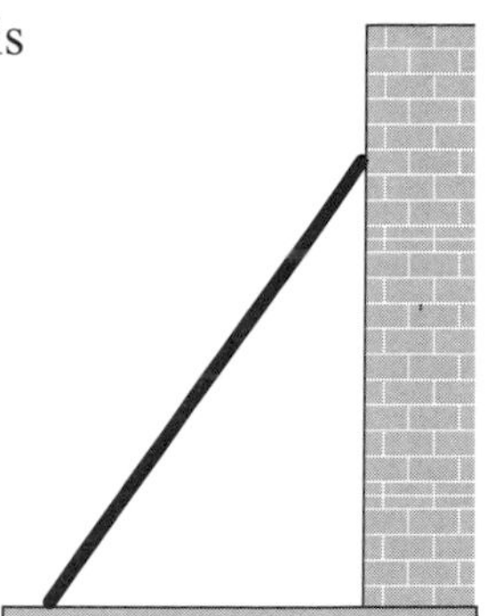

4 Meg used her 8-metre long ladder to paint her upstairs windows. She placed the top of the ladder 6 metres above the ground. How far away from the base of the wall was the foot of the ladder?

5 Susie is flying her kite on a horizontal playing field. The 300 m length of string is taut and the kite is 200 m away from Susie in a horizontal direction. How far from the ground vertically is the kite?

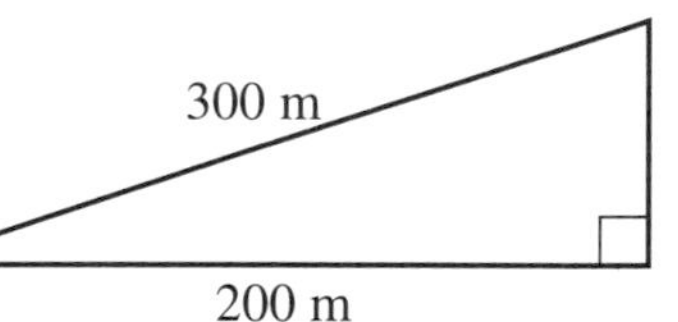

Exercise 15.3 Links: (*15D*) 15D

1 Peter and Elaine sail due south from Portsmouth for 5 miles. They then sail due west for 4 miles before turning for home. How far from Portsmouth is the boat when it turns for home?

2 Isosceles triangle ABC has two equal sides of length 8 cm and a base of length 6 cm. Calculate the height of the triangle.

3 Isosceles triangle PQR has two equal sides of length 15 cm and a height of 10 cm. Calculate the length of the base of the triangle.

4 A roof truss *XYZ* in the shape of an isosceles right-angled triangle has a span of 20 metres. Calculate the length of one of the slanted edges.

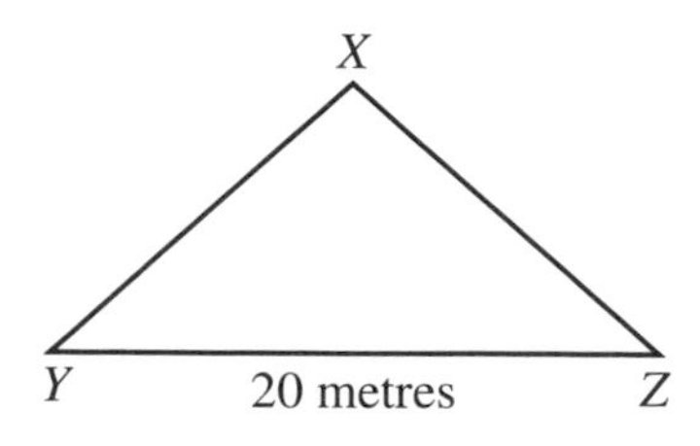

16 Averages and spread

Exercise 16.1 Links: (*16A*) 16A

1 Find the mean and median of the following:
(a) 2, 9, 1, 16, 11, 8, 3
(b) 22, 29, 21, 36, 31, 28, 23
(c) 6, 27, 3, 48, 33, 24, 9
(d) 2.6, 9.6, 1.6, 16.6, 11.6, 8.6, 3.6

2 The average rainfall in Braintree for the last five years has been 48.7 cm. How much rain must fall this year to bring the average for the six years up to 50 cm?

3 The scores for a basketball team are 72, 84, 102, 86, 73, 90 and 83. The scores in the next two matches bring the average up to 90. What was the total score for these two games?

4 The mode of the numbers 4, 6, 7, x, 3, 9, 5 is 7. Find the mean and median.

5 The median of the numbers 2, 5, 8, 10, 10, 13, y is 9. What are the mean and mode?

6 The numbers 7, 3, x, y, 12 have a mean of 5. x and y are different positive integers. Find x and y. What is the median?

7 x, x, y, 15, 12, 7, 2 have a mean of 8. x and y are integers which add up to 15. Find x and y and state the value of the median.

8 This list is in order of size. All the values are integers.

2, 6, x, 8, 13, y, z, 23, 23

The mode is NOT 23. The mean is an integer. Find x, y and z.

Exercise 16.2 Links: (*16B, C, D*) 16B, C, D

1 The numbers 11, 15, 13, 20, x, x, 12 has a mean of 15. Find x. Write down the mode, median and range.

2 The number of hours of sunshine each day for a week were recorded as:

4.5, 10.6, 5.1, 7.9, 11.2, 8.3, 10.7

Write down the median and the range.

3 23, 31, 17, y, x, 13 have a range of 20 and a mean of 18. If x is greater than y, find x and y.

4 Construct a set of five numbers that has a range of 10, a mean of 8, a median of 6 and a mode of 5.

5 The scores for 15 sets of darts by a player were

100, 60, 100, 41, 81, 140, 100, 100, 60, 53, 91, 43, 30, 100, 55

Work out the mode, median, range, upper and lower quartiles.

6 The distances, in kilometres, driven by a sales representative each day during a promotional tour were:

183, 76, 120, 34, 47, 82, 20, 150, 58, 54, 161

Work out the median and upper and lower quartiles.

7 The weights of packets of sugar are being checked to ensure that the packing machinery is working properly. The difference between the actual and stated value is recorded. Overweight is positive and underweight is negative.

+1.1 g, +2.1 g, −0.5 g, +1.3 g, +0.9 g, −0.3 g,
+1.3 g, +1.9 g, −0.4 g, +1.2 g, +2.0 g

Work out the interquartile range and the median.

Exercise 16.3 Links: (*16E, F, G*) 16E, F, G

1 The table is a record of the number of photographs taken on rolls of film which are described as 36 exposure.

Number of pictures	34	35	36	37	38
Frequency	1	3	26	31	4

Work out the mean, mode and median.

2 The bus fares in London are set according to the number of zones travelled through. The table shows the fares payable and the number of passengers purchasing these tickets.

Fare	80 p	£1.20	£2.50	£3.00
Frequency	103	72	23	12

Work out the mean, mode and median.

3 Andrew did a survey at the seaside for his science coursework. He measured 55 pieces of seaweed. The results of his survey are shown in the table.

Length of seaweed (cm)		Frequency	
$0 < L \leqslant 20$		2	
$20 < L \leqslant 40$		22	
$40 < L \leqslant 60$		13	
$60 < L \leqslant 80$		10	
$80 < L \leqslant 100$		5	
$100 < L \leqslant 120$		2	
$120 < L \leqslant 140$		1	

(a) Work out an estimate for the mean length of the pieces of seaweed. Give your answer to 1 decimal place.
(b) Write down the interval which contains the median length of a piece of seaweed. [E]

4 The grouped frequency table gives information about the weekly rainfall (d) in millimetres at Heathrow Airport in 1995.

Weekly rainfall (d) in mm		Number of weeks	
$0 \leqslant d < 10$		20	
$10 \leqslant d < 20$		18	
$20 \leqslant d < 30$		6	
$30 \leqslant d < 40$		4	
$40 \leqslant d < 50$		2	
$50 \leqslant d < 60$		2	

(a) Calculate an estimate for the mean weekly rainfall.
(b) Write down the probability that the rainfall in any week in 1995, chosen at random, was greater than or equal to 20 mm and less than 40 mm. [E]

Exercise 16.4 Links: (*16H*) 16H

1 The manager at 'Fixit Exhausts' records the time, to the nearest minute, to repair the exhausts on 20 cars. Here are his results:

32 29 34 28 22 41 57 43 28 33
35 25 52 47 39 27 36 48 53 44

Draw a stem and leaf diagram to show this information. [E]

2 The stem and leaf diagram shows the marks of some students in a test:

0	5	7					
10	0	1	1	3	6	7	
20	1	4	4	5	7	7	8
30	0	0	2	2	2		

Work out the range, the mode and the median. It is likely that the test was out of 32. Why?

3 The height of some seedlings is measured to the nearest mm. Half of the seedlings had been fed a special plant supplement. The other half had no special treatment. Here are the stem and leaf diagrams for the results:

UNTREATED							TREATED						
	7	6	6	4	3	20	2	7					
9	8	7	6	5	1	30	0	3	4				
	9	8	5	4	0	40	1	5	6	9	9		
			2	2	2	50	2	2	3	6	7	7	9
						60	0	1					

Work out the range, mode and median for the two sets. Comment on whether the treatment works.

Exercise 16.5 Links: (*16H–J*) 16I–L

1 **(a)** Draw a cumulative frequency graph for question **4** in Exercise 16.3. Use your cumulative frequency graph to estimate the median weekly rainfall. You must show your method clearly. [E]

(b) Create a box and whisker diagram for the information.

2 150 year 11 students took a mathematics examination. The table shows information about their marks.

Marks (x)	Frequency
$0 \leqslant x < 20$	6
$20 \leqslant x < 30$	17
$30 \leqslant x < 40$	22
$40 \leqslant x < 50$	45
$50 \leqslant x < 60$	26
$60 \leqslant x < 70$	19
$70 \leqslant x < 80$	9
$80 \leqslant x < 100$	6

(a) Complete a cumulative frequency table and draw a cumulative frequency diagram to show these marks.

60% of the students passed the examination.

(b) Use your diagram to find an estimate for the pass mark for the examination.

(c) Work out the interquartile range for the students' marks.

(d) Create a box and whisker diagram for the information. [E]

3 The times when students arrive at school is recorded one morning. This frequency table shows the results.

Time of arrival	Frequency
$t < 8.45$	130
$8.45 \leqslant t < 8.50$	280
$8.50 \leqslant t < 8.55$	520
$8.55 \leqslant t < 9.00$	430
$9.00 \leqslant t < 9.05$	80
$9.05 \leqslant t < 9.10$	30
$9.10 \leqslant t < 9.15$	20
$9.15 \leqslant t$	10

(a) Work out the cumulative frequencies.

(b) Construct the cumulative frequency diagram.

(c) What is the median time of arrival?

(d) Work out the interquartile range.

(e) School starts at 9.00 am. Estimate the percentage of students that are more than 7 minutes early.

(f) The last 5% of students to arrive are punished. Estimate how late these students were.

(g) Create a box and whisker diagram for the information.

4 In a golf tournament 117 competitors take part in the first round. The statistics are best score 66, worst score 85, median 74, lower quartile 70, upper quartile 77.
115 competitors take part in the second round when the statistics are best score 68, worst score 78, median 72, lower quartile 70, upper quartile 74.
The conditions for the two rounds are similar. There are two hypotheses: Players do better in the second round as they become more familiar with the course AND players do worse in the second round as they get tired. By drawing box and whisker diagrams, see what evidence there is to support or reject these hypotheses.

By the third round there are 48 competitors left. The results are best score 70, worst score 77, median 74, lower quartile 72, upper quartile 75.
Discuss these results in comparison with day two.

17 The tangent ratio

Exercise 17.1 Links: (*17A, B, C*) 17A, B, C

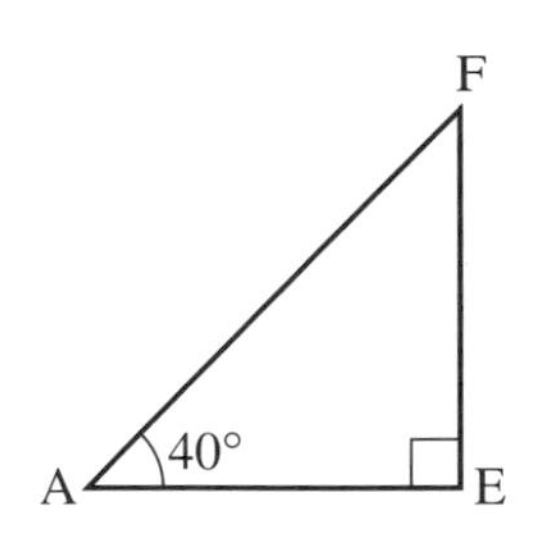

1 You need a protractor and ruler. Construct accurately triangle AEF with angle A = 40° and a right angle 90° at E. The sides can be any length.

(a) Measure the length of EF.
(b) Measure the length of AE.
(c) For the angle A (40°) work out the value of

$$\frac{\text{length of opposite side}}{\text{length of adjacent side}}$$

(d) Draw other triangles with A = 40° and E = 90° and repeat **(a)** to **(c)**.
(e) Write down what you notice about your results.

2 Use your calculator to find these tangents. Give your answers to 4 significant figures.

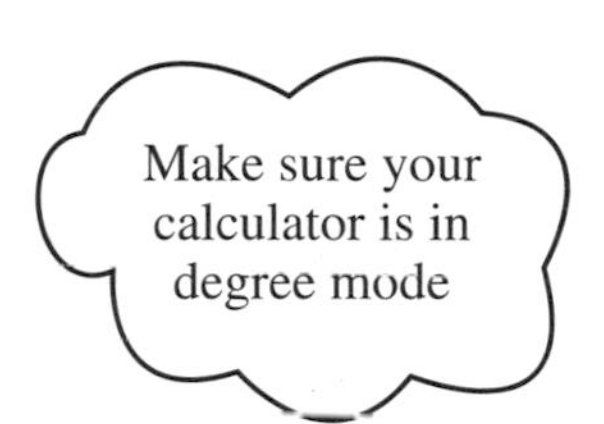

(a) tan 6° **(b)** tan 12° **(c)** tan 17°
(d) tan 51° **(e)** tan 28.5° **(f)** tan 39.2°
(g) tan 79.2° **(h)** tan 88.2° **(i)** tan 89.3°

3 Use your calculator to find the values of θ. Give your answers to 4 significant figures.

(a) $\tan\theta = 0.4$ **(b)** $\tan\theta = 93$ **(c)** $\tan\theta = 6.2$
(d) $\tan\theta = 2.7$ **(e)** $\tan\theta = 4.9$ **(f)** $\tan\theta = 5.19$

Exercise 17.2 Links: (*17D*) 17D

In this exercise give your answers correct to 1 d.p.

1 Calculate the lettered angles in these triangles:

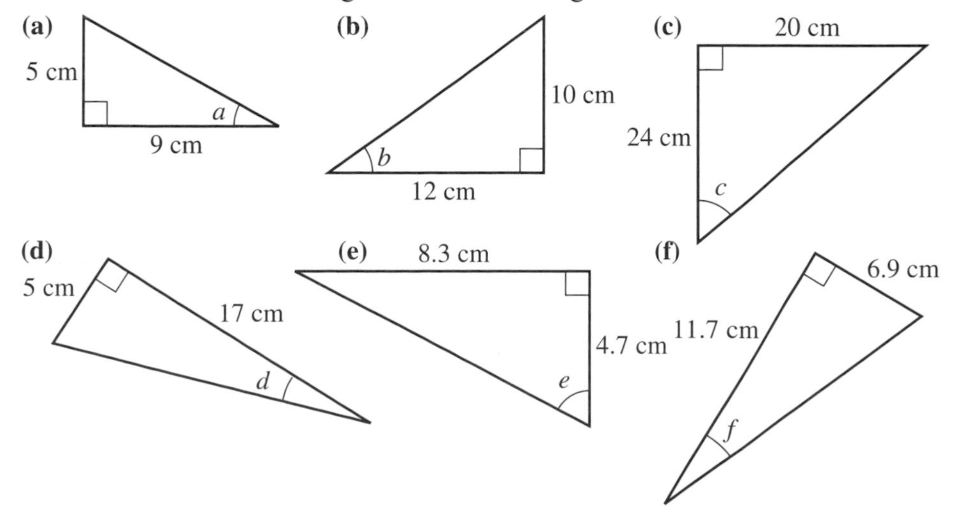

2 The diagram shows a roof strut.
The strut has a right angle at B.
The edge AB = 6 m and BC = 7 m.
Calculate the angle the roof makes with the horizontal at A.

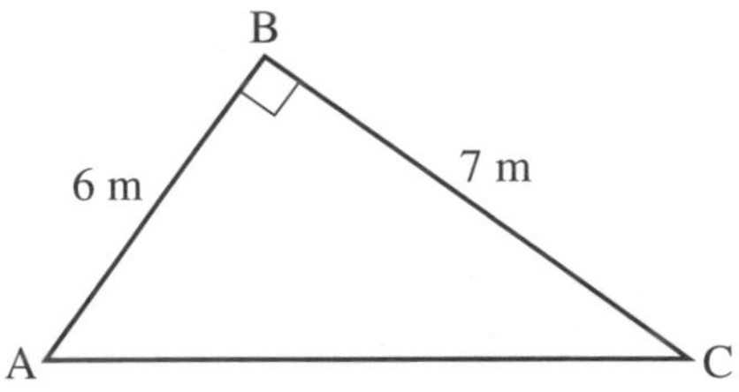

3 Leeds is 135 miles North of Colchester and 50 miles to the West.

(a) Calculate the angle marked x.

(b) Hence write down the bearing of Leeds from Colchester.

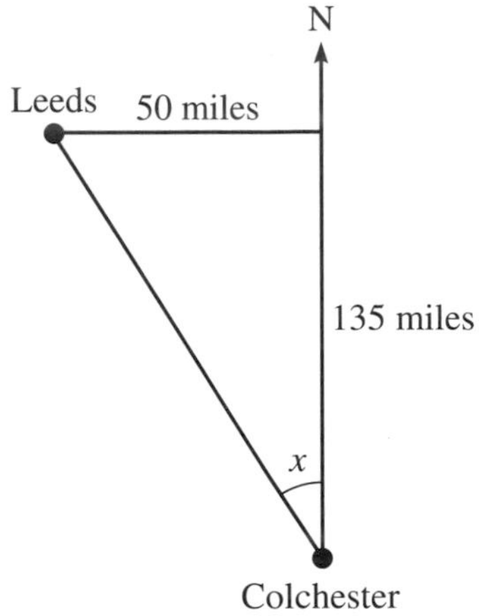

Exercise 17.3 Links: (*17E, F*) 17E, F

In this exercise give your answers correct to 3 s.f.

1 Calculate the lettered lengths in these triangles:

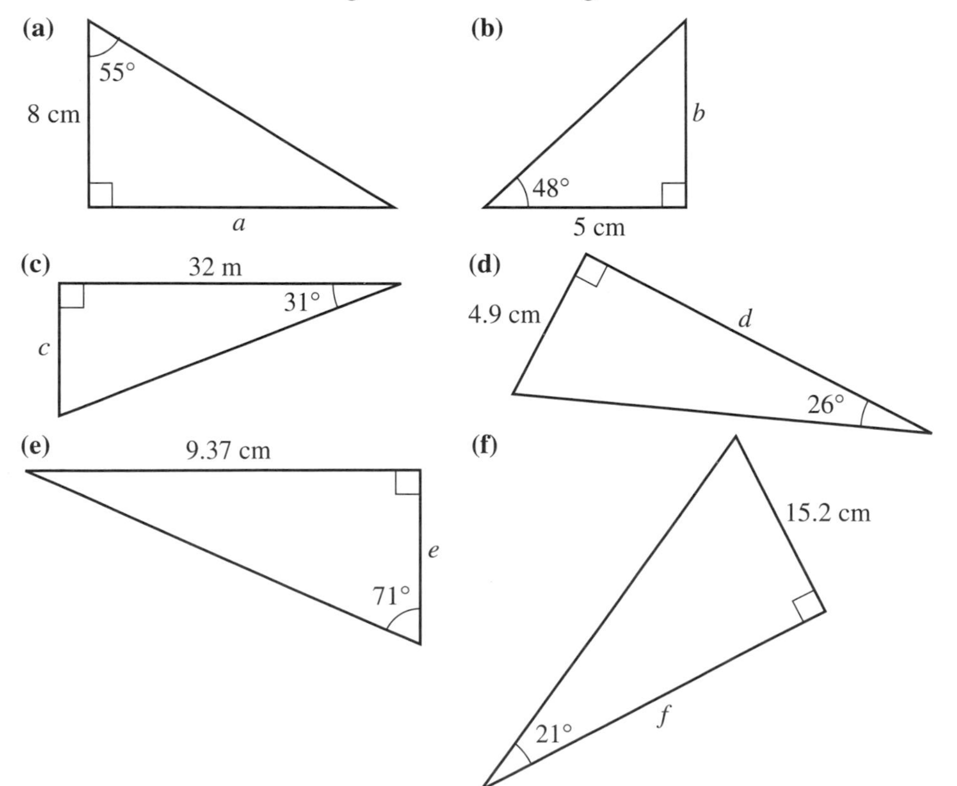

2 The length of a rectangle is 9.8 cm. The angle between a diagonal of the rectangle and the length is 30°. Calculate the width of the rectangle.

3 A guy rope from a tent makes an angle of 62° with the ground. The guy rope is 1.2 m from the tent. What is the height of the tent at this point?

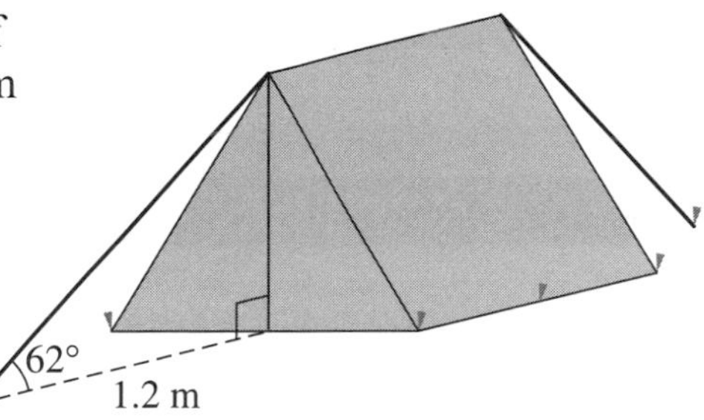

4 Calculate the named lengths in these triangles:

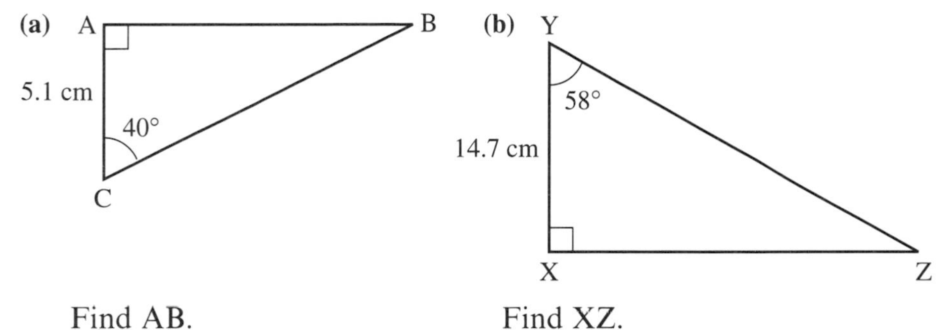

Find AB.

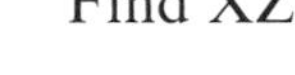

Find XZ.

5 An aircraft is descending at an angle of 4.5°. Its horizontal distance from the landing marks on the runway is 3000 m. Calculate the height of the aircraft above the ground.

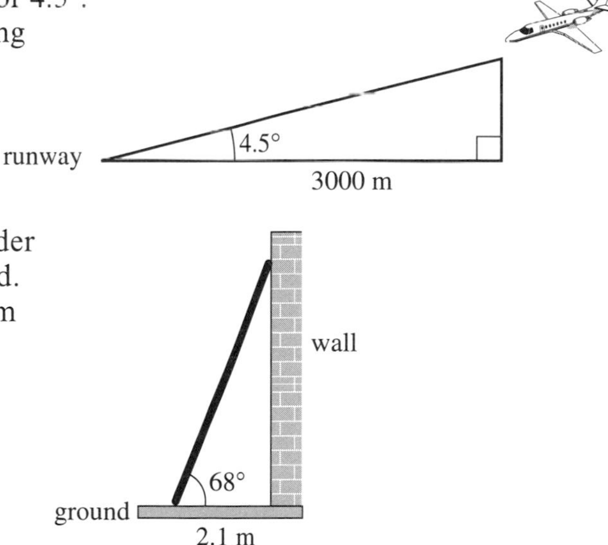

6 A ladder rests against a wall. The ladder makes an angle of 68° with the ground. The bottom of the ladder is 2.1 m from the wall. How far up the wall is the top of the ladder?

18 Graphs of more complex functions

Exercise 18.1 Links: (*18A, B, C*) 18A, B, C

You will need graph paper and tracing paper.

1 (a) Copy and complete the table of values for $y = x^2$.

x	-3	-2	-1	0	1	2	3
y		4		0	1		9

(b) Draw the graph of $y = x^2$ for values of x from -3 to 3. Label this graph **A**.

(c) On the same axes, draw and label the graphs:

B $y = x^2 + 4$ **C** $y = x^2 - 3$ **D** $y = 2x^2$

2 (a) Use tracing paper to make a copy of this graph of $y = x^2$ for values of x from 0 to 4.

(b) Complete the graph for values of x from 0 to -4.

(c) On the same axes, draw and label the graphs:

A $y = -x^2$

B $y = -x^2 + 3$

C $y = -\frac{x^2}{2}$

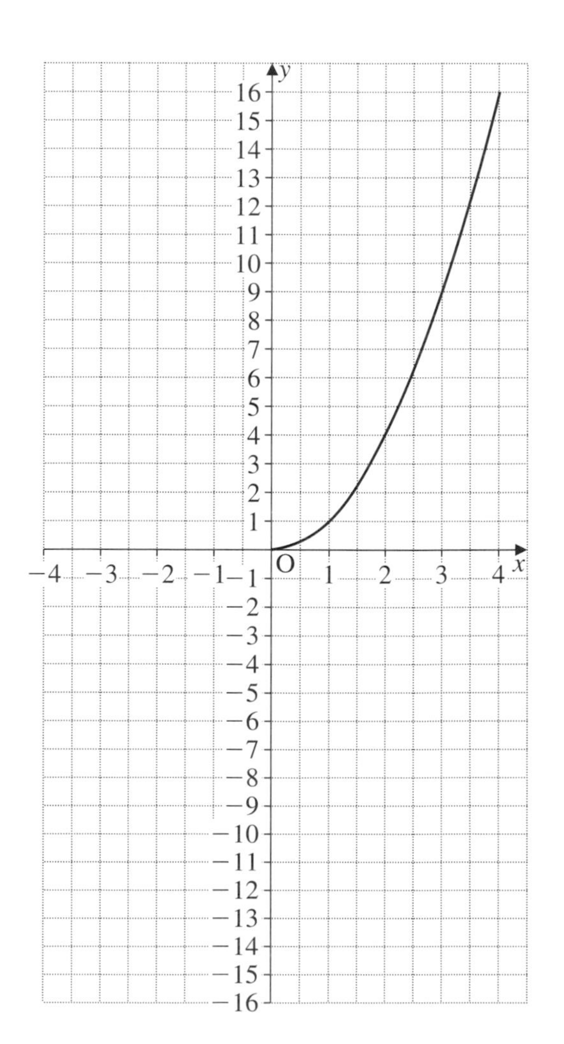

3 Here are six graphs labelled **A, B, C, D, E** and **F**:

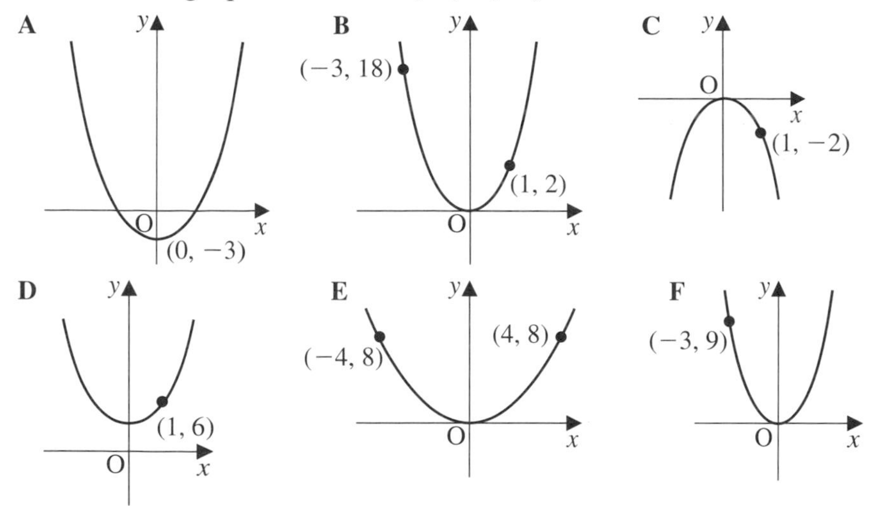

Here are six functions, labelled ①, ②, ③, ④, ⑤, ⑥.
Match each function to the correct sketch.

① $y = x^2$ ② $y = x^2 + 5$ ③ $y = x^2 - 3$

④ $y = \dfrac{x^2}{2}$ ⑤ $y = 2x^2$ ⑥ $y = -2x^2$

4 Draw the graph of $y = 2x^2 - 4$ for values of x from -3 to 3.

5 Here are the sketches of three quadratic graphs.

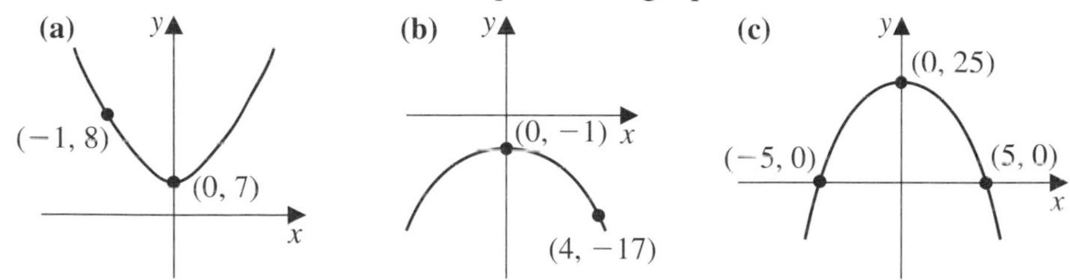

Work out the equation of each graph.

6 Solve each quadratic equation by drawing its number machine and its inverse number machine.

(a) $2x^2 - 5 = 27$ **(b)** $3x^2 + 7 = 34$ **(c)** $x^2 - 9 = 27$

(d) $5x^2 + 1 = 81$ **(e)** $3x^2 + 2 = 50$ **(f)** $\dfrac{x^2}{2} - 3 = 5$

Exercise 18.2 Links: (*18D, E, F*) 18D, E, F

1 **(a)** Copy and complete the table of values for $y = x^2 - 4x + 5$

x	−1	0	1	2	3	4	5
y		5		1			10

(b) Draw the graph of $y = x^2 - 4x + 5$ for values of x from -1 to 5.
(c) Draw the axis of symmetry for this graph.
(d) Write down the equation of this axis of symmetry.

2 **(a)** Copy and complete the table of values for $y = 12 - x^2$

x	-3	-2	-1	0	1	2	3
y		8			11		3

(b) Draw the graph of $y = 12 - x^2$ for values of x from -3 to 3.
(c) Draw the axis of symmetry for this graph.
(d) Write down the equation of this axis of symmetry.

3 Draw the graphs of each of the following:
(a) $y = x^2 - 3$ for values of x from -3 to 3
(b) $y = 15 - x^2$ for values of x from -4 to 4
(c) $y = x^2 - 4x$ for values of x from -1 to 5
(d) $y = x^2 + 6x$ for values of x from -7 to 1
(e) $y = x^2 - 2x - 3$ for values of x from -3 to 5
(f) $y = 2x^2 + 4x - 1$ for values of x from -4 to 2

In each case, draw the axis of symmetry, and write down the equation of the axis of symmetry.

4 Work out the equation of the axis of symmetry for each of these quadratic graphs:
(a) $y = x^2 + 7$ **(b)** $y = -2x^2$ **(c)** $y = x^2 - 4x + 3$
(d) $y = 3x^2 - 12x + 8$ **(e)** $y = -x^2 + 5x + 3$

5 **(a)** Draw the graph of $y = 2x^2 - 8x + 5$ for values of x from -1 to 5.
(b) Find the minimum value of y and the value of x when this minimum occurs.

6 Given that $y = -x^2 + 4x + 3$, find the maximum value of y.

7 **(a)** Copy and complete the table of values for $y = 2x^2 - 4x + 7$

x	-2	-1	0	1	2	3	4
y		13					23

(b) Use the graph to solve each of the following equations:
(i) $2x^2 - 4x + 7 = 13$
(ii) $2x^2 - 4x + 7 = 16$
(c) State clearly why there are **no solutions** to the equation $2x^2 - 4x + 7 = 0$.
(d) Find:
(i) the minimum value of y
(ii) the corresponding value of x.

8 The diagram shows a rectangle PQRS.
PQ = x cm
The perimeter of PQRS is 16 cm.

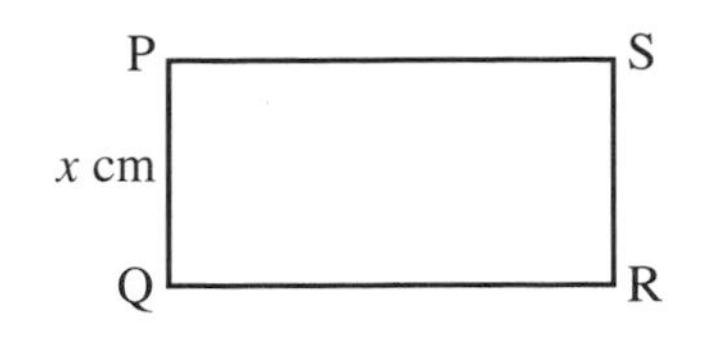

(a) Write down, in terms of x, an expression for the length of QR in centimetres.

The area of PQRS is A cm^2

(b) Show that $A = 8x - x^2$

(c) Copy and complete the table of values below:

x (cm)	0	1	2	3	4	5	6	7	8
A (cm^2)	0					15			0

(d) Draw the graph of A against x for values of x from 0 to 8.

(e) Use your graph to find:
  (i) the maximum value of A
  (ii) the value of x for which A is a maximum.

(f) Use your graph to find:
  (i) the value of A when $x = 1.5$ cm
  (ii) the values of x when $A = 13.4$ cm^2.

Exercise 18.3 Links: (*18G, H*) 18G, H

1 On the same axes, draw and label with the appropriate letters, the graphs of:

(a) $y = x^3$ for values of x from −3 to 3
(b) $y = x^3 + 4$ for values of x from −3 to 3
(c) $y = 2x^3$ for values of x from −2 to 2
(d) $y = \frac{x^3}{3} - 1$ for values of x from −4 to 4
(e) $y = -x^3 + 3$ for values of x from −3 to 3

2 On the same axes, draw and label with the appropriate letter, the graphs of:

(a) $y = \frac{1}{x}$ for values of x from −4 to 4
(b) $y = \frac{12}{x}$ for values of x from −6 to 6
(c) $y = \frac{3}{x} + 2$ for values of x from −3 to 3
(d) $y = \frac{-4}{x} + 1$ for values of x from −4 to 4
(e) $y = \frac{-6}{x} - 2$ for values of x from −6 to 6

3 Sketches of six graphs are shown below.
These sketches are labelled **A, B, C, D, E, F**

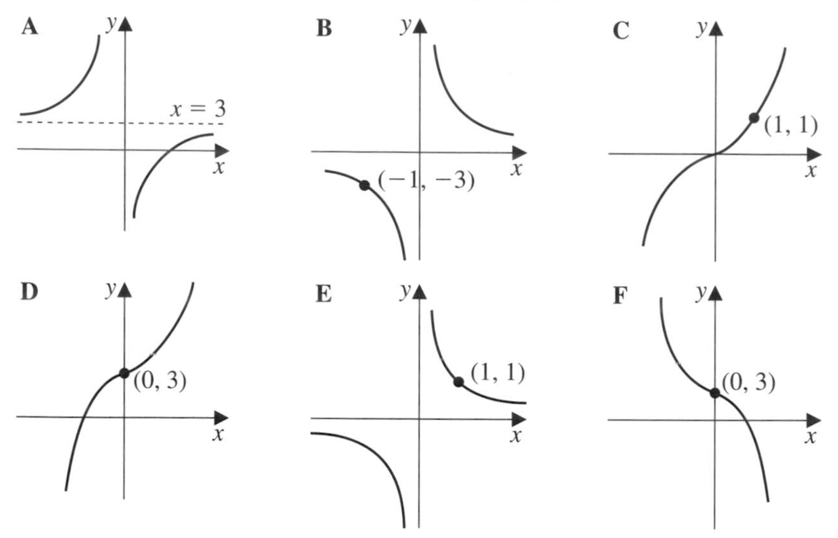

Here are six functions, labelled ①, ②, ③, ④, ⑤, ⑥.
Match each function to the correct sketch.

① $y = x^3$ ② $y = \frac{1}{x}$ ③ $y = \frac{-1}{x} + 3$

④ $y = -x^3 + 3$ ⑤ $y = \frac{3}{x}$ ⑥ $y = 2x^3 + 3$

4 **Without working out a table of values**, sketch the graph for each of these functions:

(a) $y = x^3$ **(b)** $y = \frac{1}{x}$ **(c)** $y = x^3 - 4$

(d) $y = \frac{1}{x} + 2$ **(e)** $y = \frac{-6}{x} + 5$ **(f)** $y = 2x^3 + 5$

5 **(a)** Copy and complete the table of values for $y = x^3 + x^2 - 3$

x	−3	−2	−1	0	1	2	3
y		−7					33

(b) Draw the graph of $y = x^3 + x^2 - 3$ for values of x from −3 to 3.
(c) Use your graph to solve the equation $x^3 + x^2 - 3 = 0$.

6 **(a)** Copy and complete the table of values for $y = x + \frac{12}{x}$

x	−6	−5	−4	−3	−2	−1	−0.5	0.5	1	2	3	4	5	6
y	−8	−7.4												8

(b) Draw the graph of $y = x + \frac{12}{x}$

(c) Use your graph to solve the equation $x + \frac{12}{x} = 9$ for values of x between −6 and 6.

Exercise 18.4 Links: (*18I*) 18I

1 **(a)** Show that the equation $x^3 - 5x = 32$ has a solution between $x = 3$ and $x = 4$.
(b) Use a method of trial and improvement to find this solution correct to one decimal place.

2 **(a)** Show that the equation $x^3 + 3x^2 = 30$ has a solution between $x = 2$ and $x = 3$.
(b) Use a method of trial and improvement to find this solution correct to two places of decimals.

3 **(a)** Sketch the graph of $y = x^2 + \frac{12}{x}$ for values of x from 1 to 6.
(b) Use your graph to find an approximate solution to the equation $x^2 + \frac{12}{x} = 30$
(c) Use a method of trial and improvement to find this solution correct to two places of decimals.

4 Use a method of trial and improvement to find, correct to 1 d.p. the positive solution of $x^3 + 3x = 40$.

Exercise 18.5 Links: (*18J, K*) 18J, K

1 Here are the names of five friends, with their weights and heights:

Name	Weight (kg)	Height (m)
Angel	44	1.57
Barry	81	1.73
Carmen	62	1.60
Derek	54	1.72
Edwina	58	1.68

Using the first letter of each person's name as a label, plot these as points on a graph.

2 Katrina boiled an egg. The sketch graph below is a graph of the temperature of the water against time during the boiling process.

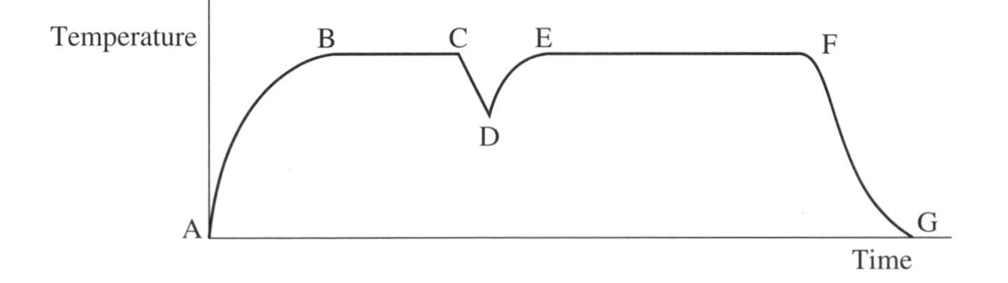

(a) Describe fully how the temperature of the water changed between the points marked A and B on the graph.
(b) What do you think happened at the time corresponding to point C on the graph?
(c) Describe fully how the temperature of the water changed between the points E, F and G on the graph, stating clearly the real-life situation you believe is represented by this portion of the graph.

3 The table below shows information connecting the speed of a car to the shortest distance in which it can stop.

Speed (mph)	Shortest stopping distance (feet)
20	40
30	75
40	120
50	175
60	240
70	315

(a) Plot the graph of shortest stopping distance against speed.
(b) Use your graph to estimate:
 (i) the shortest stopping distance for a car travelling at 45 mph.
 (ii) the speed of a car when its shortest stopping distance is 300 feet. [E]

4

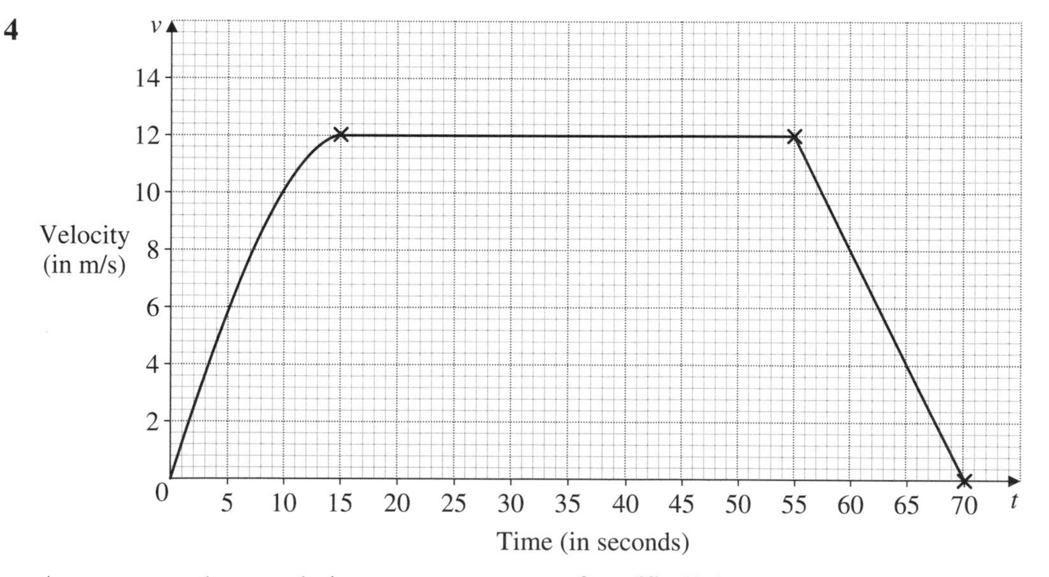

A motor cycle travels between two sets of traffic lights.
The diagram is the velocity/time graph of the motor cycle.

The motor cycle leaves the first set of traffic lights.
(a) Use the graph to find the velocity of the motor cycle after 30 seconds.
(b) Describe fully the journey of the motor cycle between the two sets of traffic lights.

19 Probability 2

Exercise 19.1 Links: (*19A, B, C, D, E*) 19A, B, C

1 Peter goes to work by bus, train or taxi. Draw a space diagram for the possibilities of Peter's travel on two consecutive days. Draw a tree diagram for the possibilities.

2 Pat goes to work by car. On her journey there is a level crossing, a set of traffic lights and a roundabout. At each of these he is either stopped or allowed to continue. Draw a space diagram and a tree diagram to show all the possibilities on one journey to work.

3 A darts player at practice records that he was successful in throwing a 'treble twenty' 63 times in 262 throws.
(a) Calculate the probability that he will be successful on a single throw.
(b) Work out an estimate for the number of times he will be successful with 400 throws.

4 You need 10 drawing pins. Throw them on to a flat surface. Record the number with points uppermost. Repeat this 20 times. Use your results to obtain an estimate for the probability of a drawing pin landing on its base.

5 Alison, Brenda, Claire and Donna are runners in a race. The probabilities of Alison, Brenda, Claire and Donna winning the race are shown below:

Alison	Brenda	Claire	Donna
0.31	0.28	0.24	0.17

(a) Calculate the probability that either Alison or Claire will win the race.

Hannah and Tracy play each other in a tennis match. The probability of Hannah winning the tennis match is 0.47.
(b) Complete a probability tree diagram. [E]

6 A fair dice is to be thrown.
(a) Write down the probability of the dice landing on:
(i) a six
(ii) an even number

A second dice is to be thrown. The probability that this dice will land on each of the numbers 1 to 6 is given in the table:

number	1	2	3	4	5	6
probability	x	0.2	0.1	0.3	0.1	0.2

The dice is to be thrown once.

(b) Calculate the value of x.

(c) Calculate the probability that the dice will land on a number higher than 3.

The dice is thrown 1000 times.

(d) Estimate the number of times the dice is likely to land on a six. [E]

7 All female chaffinches have the same pattern of laying eggs. The probability that any female chaffinch will lay a certain number of eggs is given in the table below:

Number of eggs	0	1	2	3	4 or more
Probability	0.1	0.3	0.3	0.2	x

(a) Calculate the value of x.

(b) Calculate the probability that a female chaffinch will lay less than 3 eggs. [E]

20 Lengths, areas and volumes

Exercise 20.1 Links: (*20A*) 20A

1 Work out the perimeters of these shapes:

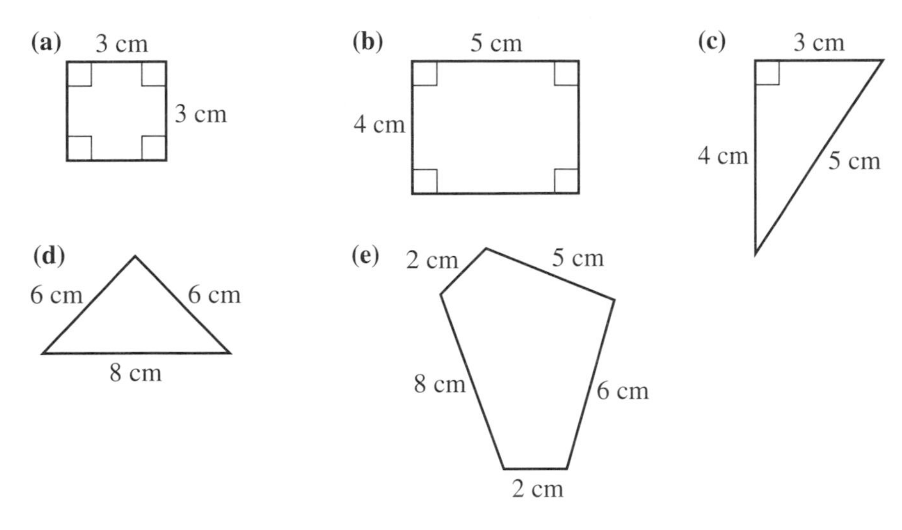

2 Find the perimeters of these shapes.

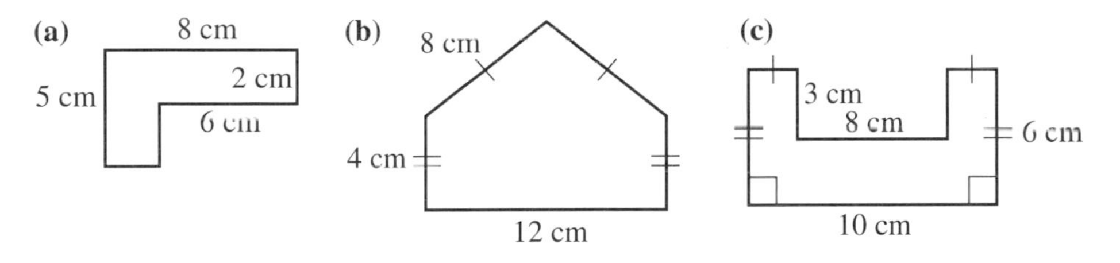

3 Find the areas of these shapes.

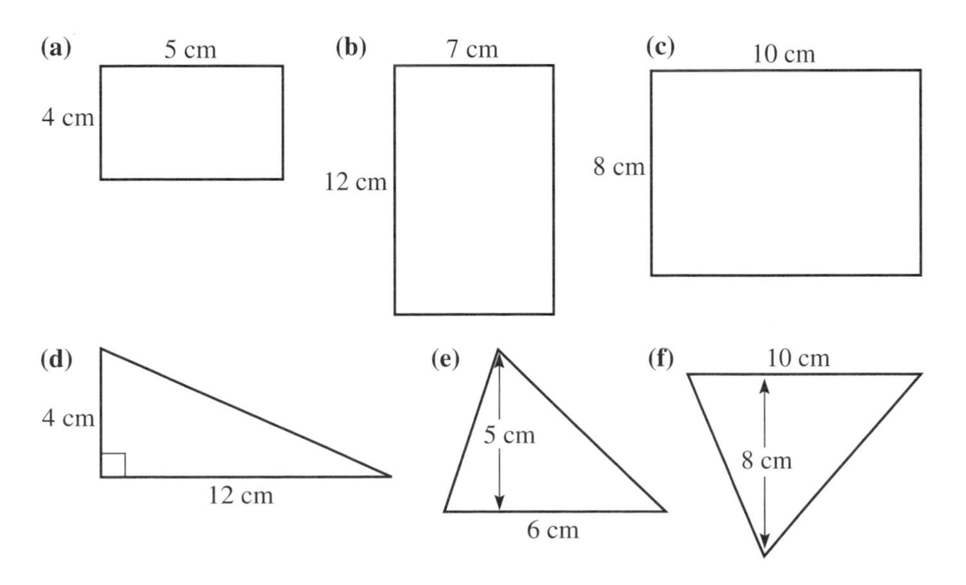

4 Find the areas of these shapes.

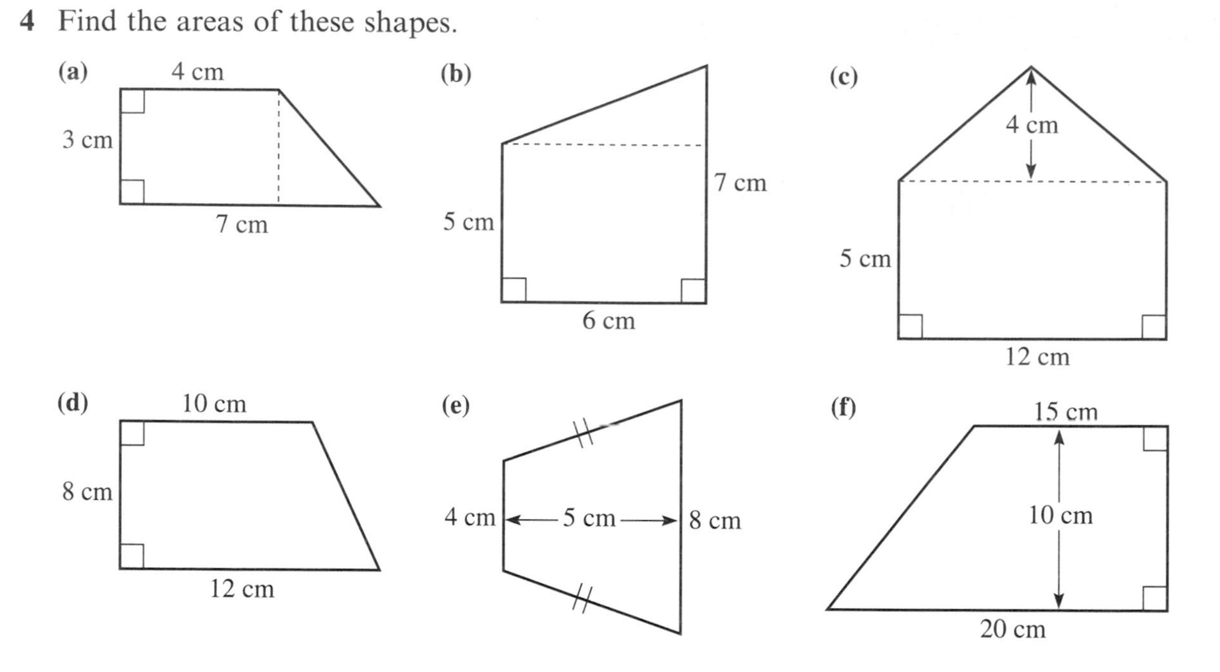

5 Copy this table into your book and complete the missing numbers.

Shape	Length	Width	Area
(a) Rectangle	4 cm	5 cm	
(b) Rectangle	5 cm	8 cm	
(c) Rectangle	8 cm		32 cm^2
(d) Rectangle	7 cm		28 cm^2
(e) Rectangle		2 cm	16 cm^2
(f) Rectangle		9 cm	108 cm^2

6 Copy this table into your book and complete the missing numbers.

Shape	Base	Vertical height	Area
(a) Triangle	6 cm	5 cm	
(b) Parallelogram	8 cm	6 cm	
(c) Triangle	8 cm		16 cm^2
(d) Triangle	7 cm		21 cm^2
(e) Parallelogram		6 cm	42 cm^2
(f) Triangle		2 cm	12 cm^2
(g) Triangle		9 cm	36 cm^2
(h) Parallelogram	7.5 cm		30 cm^2

Exercise 20.2 Links: (*20C, E*) 20C, E, F

1 Work out the circumferences of these circles with diameters:
(a) 5 cm **(b)** 4 cm **(c)** 6 cm **(d)** 10 cm
(e) 1.5 m **(f)** 2.8 m **(g)** 3.6 m **(h)** 12 m
(i) 25 cm **(j)** 1 km **(k)** 2.25 m **(l)** 5.6 cm

2 Work out the circumferences of these circles with radii:
(a) 5 m **(b)** 3 cm **(c)** 2 m **(d)** 2.5 m
(e) 3.5 cm **(f)** 4.5 m **(g)** 2.4 m **(h)** 5.6 cm
(i) 20 mm **(j)** 30 mm **(k)** 2.1 m **(l)** 1 m

3 Find the areas of the circles with radii:
(a) 5 cm **(b)** 7 cm **(c)** 11 cm **(d)** 4 cm
(e) 2.5 cm **(f)** 3.2 cm **(g)** 5.4 m **(h)** 2.9 m

4 Find the areas of the circles with diameters:
(a) 8 cm **(b)** 6 cm **(c)** 10 cm **(d)** 18 cm
(e) 3.2 cm **(f)** 8.4 cm **(g)** 6.6 m **(h)** 12.4 m

5 A wheelbarrow has a wheel with a diameter of 30 cm. Work out the circumference of the wheelbarrow wheel.

6 A bicycle has a wheel with a radius of 25 cm. Work out the circumference of the bicycle wheel.

7 Work out the area of a:
(a) circular pond with a radius of 0.9 m
(b) circular birthday card with a diameter of 120 mm
(c) crop circle with a radius of 15 m
(d) pencil with a diameter of 1 cm
(e) circular mill wheel with a radius of 30 cm.

8 Robin travels 600 metres on a bicycle. The circumference of the bicycle wheel is 75 cm. How many times does the wheel rotate on the journey?

9 Sylvie travels 1.2 kilometres on her bicycle. The diameter of the bicycle wheel is 60 cm.
(a) Work out the circumference of the bicycle wheel.
(b) How many times does the wheel rotate on the journey?

10 A circle has an area of 10 cm^2. Work out the radius.

11 A tree with a circular trunk has a circumference of 2.8 m. Work out the diameter of the tree trunk.

12 Copy this table into your book. Work out the diameter and radius of the circles.

Circumference	Diameter	Radius
(a) 50 cm		
(b) 45 m		
(c) 10 mm		
(d) 15 cm		
(e) 3.14 m		

13 One of Sven's rollerblade wheels rotates 5000 times when he travels 600 metres. Work out the diameter of the wheel.

14 The circumference of a circular village pond is 60 metres. Work out the radius.

15 A circular table top has an area of $3\,\text{m}^2$. Work out the diameter of the table.

16 Work out in terms of π the area and circumference of the circles with:
(a) radius
  (i) 3 cm **(ii)** 10 cm **(iii)** 20 mm
(b) diameter
  (i) 6 cm **(ii)** 5 m **(iii)** 10 mm

17 A table top in the shape of a circle has a diameter of 90 cm.
(a) Work out the circumference of the table top in terms of π.
(b) Work out the area of the table top in terms of π.

90 cm

18 The area of a circle is $100\pi\,\text{cm}^2$.
Calculate: **(a)** the radius
(b) the circumference of the circle.

Exercise 20.3 Links: (*20F*) 20G

1 For these cuboids, work out:
(i) their volumes **(ii)** their surface area.

(a)

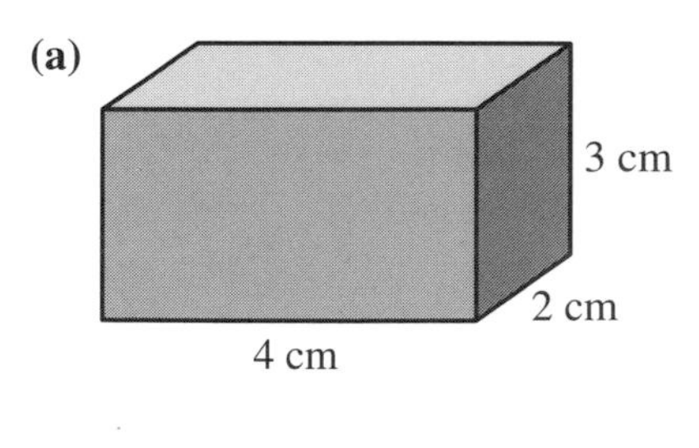

(b)

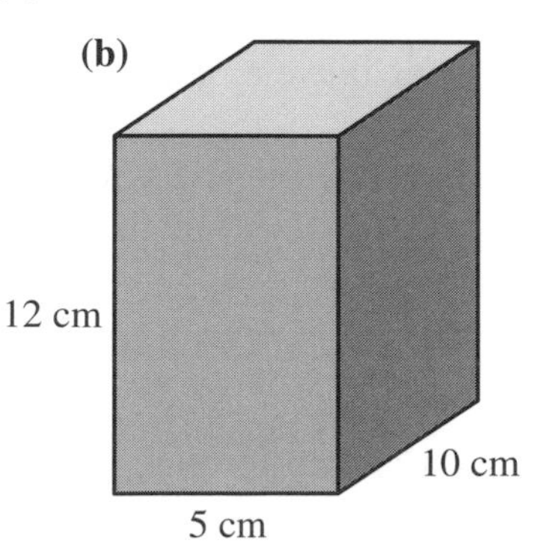

(c)

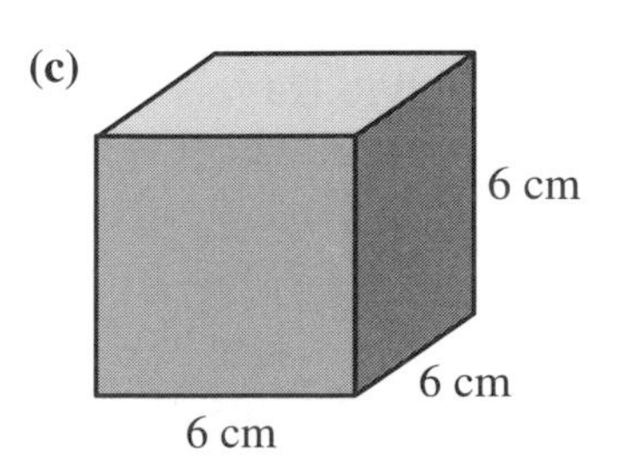

2 How many cubes with edge length 2 cm will fit into a cuboid measuring 10 cm by 8 cm by 6 cm?

3 Work out the volumes of these prisms.

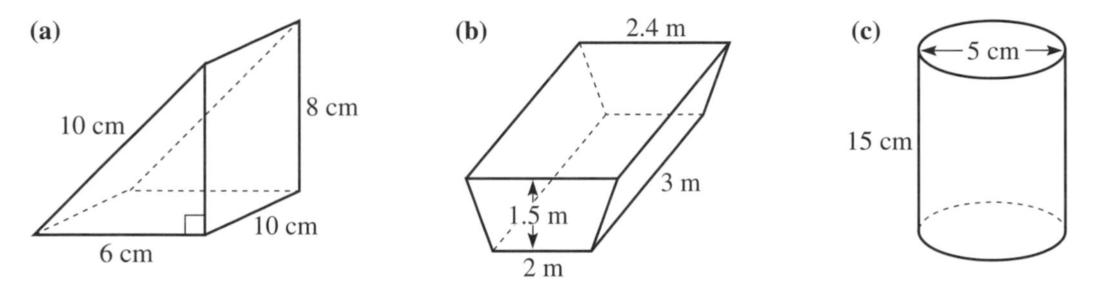

4 Copy the table into your book and fill in the missing numbers.

Shape	Length	Width	Height	Volume	Surface area
(a) Cuboid	6 cm	4 cm	2 cm		
(b) Cube	3 cm				
(c) Cuboid	6 cm		3 cm	36 cm^3	
(d) Cuboid	8 cm		2 cm	48 cm^3	
(e) Cuboid		4 cm	2 cm	24 cm^3	
(f) Cuboid		5 cm	3 cm	75 cm^3	
(g) Cuboid	4 cm	2 cm		32 cm^3	
(h) Cube					600 cm^2

5 Calculate the surface area of the prisms in question **3**.

6 A tin of sweets in the shape of a cylinder with radius 12 cm and height 12 cm has a paper wrapper covering the curved surface. Calculate the area of the paper wrapper. You may assume that there is no overlap.

7 The body of this wheelbarrow has the shape of a prism with a cross-section that is a trapezium. The length of the top is 1.2 m and the length of the bottom is 40 cm. It has a depth of 35 cm and a width of 80 cm. Work out how many barrow-loads it would take to fill a skip with a capacity of 8 m^3. (You may assume that a barrow-load is the volume of the barrow.)

8 How many cylindrical glasses with a height of 150 mm and a diameter of 50 mm can be filled from a 3-litre bottle of cola?

Exercise 20.4 Links: (*20G*) 20H

In this exercise all the letters represent lengths; π and all numbers have no dimension.

1 For each of these expressions write down whether it represents a length, an area or a volume.

(a) πD **(b)** πr^2 **(c)** $4\pi r^3$ **(d)** $2(l + w)$
(e) h^3 **(f)** $k \times g$ **(g)** πrl **(h)** $3x^2y$

2 For each of these expressions explain why they cannot represent a length, an area or a volume.

(a) $ab + c$ **(b)** $abcd$ **(c)** b^2c^3 **(d)** $\frac{3abc}{abc}$

3 Tick ✓ the box that applies for each of these expressions:

Expression	$\pi(a+b)$	πr^3	pqr	a^3b^2	xy	$ab + ac$	$3b(a+c)$	$\pi(r^2 - r^3)$
Length								
Area								
Volume								
None of these								

21 Algebraic expressions and formulae

Exercise 21.1 Links: (*21A, B*) 21A, B

1 Work out the value of these algebraic expressions using the values given:

(a) $5a + 3$ if $a = 4$
(b) $4b - c$ if $b = 2, c = 5$
(c) $3p - 2q$ if $p = 5, q = 2$
(d) $xy - z$ if $x = 2, y = 4, z = 3$
(e) $12 + 5t$ if $t = -2$
(f) $p - 3t$ if $p = 4, t = -2$
(g) $4y + 7$ if $y = 4\frac{1}{2}$
(h) $6st$ if $s = \frac{1}{2}, t = \frac{3}{4}$
(i) $4(a + b)$ if $a = 2, b = 3$
(j) $5(x - y)$ if $x = 7, y = 4$
(k) $x(6 - y)$ if $x = 3, y = 2$
(l) $3(8 - t)$ if $t = -2$
(m) $\frac{1}{2}(a + b)$ if $a = 3, b = 5$
(n) $-2(3t + 1)$ if $t = -2$
(o) $3(2x - y)$ if $x = -1, y = -3$
(p) $4(p + 2q)$ if $p = 1, q = -2\frac{1}{2}$
(q) $6(a + b)$ if $a = 2, b = -2$
(r) $\frac{3}{4}(f + g)$ if $f = 2, g = -10$

2 Work out the values of each of these expressions for the given values of the letters:

(a) $\frac{x}{5} + 2$ if $x = 15$
(b) $\frac{p}{q} - 2$ if $p = 36, q = 3$
(c) $\frac{n - 3}{5}$ if $n = 13$
(d) $\frac{a + 2b}{3}$ if $a = 1, b = 4$
(e) $\frac{ab + c}{2}$ if $a = 2, b = 3, c = 4$
(f) $\frac{2pq}{12} - 3$ if $p = 4, q = 6$
(g) $\frac{4xy - z}{7}$ if $x = 2, y = 3, z = -4$
(h) $\frac{3t}{15} + s$ if $t = 5, s = -1$
(i) $2t - \frac{3r}{4}$ if $t = 5, r = 8$
(j) $a - \frac{4b}{6}$ if $a = -2, b = -12$

3 Work out the value of each of these algebraic expressions using the values given:

(a) $3t^2 + 2$ if $t = 4$
(b) $2n^2 - 3$ if $n = 5$
(c) $4x^2$ if $x = \frac{1}{2}$
(d) $8y^2 + 1$ if $y = -1$
(e) $5 - p^2$ if $p = -3$
(f) $3a^2 + 2b^2$ if $a = -1, b = 3$
(g) $x^2 + 2x$ if $x = 4$
(h) $3t^2 - 2t$ if $t = 5$
(i) $x^2 - y^2$ if $x = 10, y = 3$
(j) $s^2 - t^2$ if $s = -8, t = 6$
(k) $\frac{1}{2}(x + x^2)$ if $x = 5$
(l) $mn + m + n$ if $m = n = 4$
(m) $\frac{x^2}{2} - 3$ if $x = 5$
(n) πr^2 if $\pi = \frac{22}{7}, r = 7$

(o) $\frac{a^2}{3} - \frac{b^2}{8}$ if $a = 6, b = 4$ **(p)** $3a(x^2 + y^2)$ if $a = 2, x = 3, y = 4$

(q) $\frac{2v^2 + u^2}{10}$ if $v = 5, u = 3$ **(r)** $\frac{1}{2}(a^2 - b)$ if $a = -4, b = -6$

(s) $\frac{a - 3b^2}{4}$ if $a = -1, b = 1$ **(t)** $\frac{x^2 - 3y^2}{4z^2}$ if $x = -5, y = -3, z = -\frac{1}{2}$

Exercise 21.2 Links: (*21C–F*) 21C–F

1 Simplify:

(a) $x^3 \times x^2$ **(b)** $y^4 \times y^5$ **(c)** $a^3 \times a^5$
(d) $a \times a^2$ **(e)** $(b^2)^3$ **(f)** $x^6 \div x^2$
(g) $b^8 \div b^3$ **(h)** $c^{20} \div c^{14}$ **(i)** $n^2 \div n$
(j) $(b^4)^5$ **(k)** $(a^3)^0$ **(l)** $x^3 \times x^2 \times x$
(m) $4x^3 \times 3x^2$ **(n)** $5p \times 2p^3$ **(o)** $2y^{10} \times y^3$
(p) $6x^3 \div 2x$ **(q)** $24p^7 \div 8p^3$ **(r)** $12a^6 \div 3a^2$
(s) $(3a^2)^3$ **(t)** $(5b^7)^2$ **(u)** $\frac{4x^3 \times 6x^2}{3x^4}$
(v) $\frac{a^4 \times a^3}{a^2}$ **(w)** $\frac{3b^3 \times 4b}{2b^4}$ **(x)** $\frac{(4a^5)^2}{8a^3}$

2 Expand the brackets:

(a) $3(x + 2)$ **(b)** $5(2p + 1)$ **(c)** $4(x - 3)$
(d) $2(3p - 2)$ **(e)** $3(a + 2b)$ **(f)** $9(2n - 5)$
(g) $x(2x + 5)$ **(h)** $a(x + a)$ **(i)** $p(3p + 1)$
(j) $3x(2x - 1)$ **(k)** $n(n + 2)$ **(l)** $4b(b + 2)$
(m) $2t^2(t + 3)$ **(n)** $5x(2 - x)$ **(o)** $4y^3(2y - 3y^2)$

3 Simplify each of these expressions:

(a) $3(2x + 1) + x$ **(b)** $5(2a + 1) - 3a$
(c) $2(x + 3y) + x + y$ **(d)** $3(3p - q) + 4(p + 2q)$
(e) $4(3a - 2b) + 2(a + b)$ **(f)** $x(x - 2) + x$
(g) $y(2y - 3) - 5y$ **(h)** $5n(2 + 3n) - 6n$
(i) $m(m^2 + m) - 2m^2$ **(j)** $4t^2(t - 1) - t^3$
(k) $\frac{1}{2}(2x^2 + 4x) + x^2 - x$ **(l)** $3n^2(n^3 + 2n) + n(n^2 - 3n)$
(m) $5x(2x + x^3) - x^4$ **(n)** $y^2(3y^2 + 5y) + 2y^3(1 - 3y)$

4 Simplify each of these expressions:

(a) $\frac{3n + 12}{3}$ **(b)** $\frac{15x - 10}{5}$ **(c)** $\frac{2a(3a + 6)}{2}$
(d) $\frac{6x^2 - 12x}{3}$ **(e)** $\frac{8r + 20s}{4}$ **(f)** $\frac{12a + 6b + 9c}{3}$
(g) $\frac{49x - 42y}{7}$ **(h)** $\frac{6a^2 + 4a^2}{5a}$ **(i)** $\frac{5(6t + 8s)}{10}$

5 Simplify each of these expressions:

(a) $5x - 2(x - y)$
(b) $8t - 3(2t - 1)$
(c) $3(2n + m) - 5(n - m)$
(d) $2(3x - 2) - 4(x - 1)$
(e) $4(3 - x) - (3 + x)$
(f) $a(a - b) - a(a + b)$
(g) $6x(x + 3) - x(6 + x)$
(h) $\frac{1}{2}(4x^2 + 6) - (1 - x^2)$
(i) $\frac{4x + 6}{2} - (x - 2)$
(j) $5y - 1 - \frac{6y - 2}{2}$
(k) $3(2 - a) - \frac{4a - 10}{2}$
(l) $p(p - 1) - p(1 - p)$
(m) $x(3x - 4) - 2(2x^2 - 5)$
(n) $-3(b - a) - (a + b)$
(o) $p^3\left(3 - \frac{1}{p}\right) - \frac{1}{2}(p^2 - 2)$
(p) $x(x - y) - y(x + y)$
(q) $x(5x + 4) - 2x(x - 3)$
(r) $n^2(1 - n) - n(1 + n)$
(s) $y^3(1 + y) - y(3 + y^3)$
(t) $8t(2 - \frac{1}{2}t) + 3(1 + t^2)$
(u) $3x - 3(x - y)$
(v) $4a(1 - a^2) - 3(a + a^2)$

Exercise 21.3 Links: (*21G*) 21G

1 Multiply out the brackets and, where possible, simplify:

(a) $(x + 5)(y + 3)$
(b) $(x + 3)(x + 5)$
(c) $(y + 2)(y + 7)$
(d) $(a + 4)(b + 2)$
(e) $(n + 1)(m + 1)$
(f) $(p + 2)(p + 4)$
(g) $(2a + 1)(a + 1)$
(h) $(b + 3)(2b + 4)$
(i) $(2p + 5)(p + 2)$
(j) $(2a + 3)(3b + 2)$
(k) $(x - 4)(x - 5)$
(l) $(y - 2)(y - 5)$
(m) $(a - 5)(a - 7)$
(n) $(2p - 3)(p - 2)$
(o) $(3q - 5)(4q - 1)$
(p) $(x - 1)(x - 2)$
(q) $(a - 2)(b - 3)$
(r) $(3z - 4)(4z - 3)$
(s) $(5x - 2)(2x - 5)$
(t) $(x - 2)(x + 2)$
(u) $(y - 3)(y + 3)$
(v) $(a + 5)(a - 5)$
(w) $(3x - 2)(x + 7)$
(x) $(2a + 1)(a - 4)$
(y) $(2b + 5)(3b - 2)$
(z) $(5n + 7)(3n - 4)$

2 Multiply out the brackets and simplify:

(a) $(p + 2q)(q + 2p)$
(b) $(x + y)(x - y)$
(c) $(x - y)(x + 2y)$
(d) $(n - 2m)(n + 3m)$
(e) $(a + 3b)(3a - b)$
(f) $(2x + 3y)(3x + 4y)$
(g) $(4a - 1)(a + 2)$
(h) $(3x - 2)(4x + 3)$
(i) $(x + 5)^2$
(j) $(y + 1)^2$
(k) $(2a + 3)^2$
(l) $(4x - 3)^2$
(m) $(2 - x)^2$
(n) $(3 - 5y)^2$
(o) $(a - 3)^2$
(p) $(x + y)^2$
(q) $(3x + 2y)^2$
(r) $(4p - 5y)^2$
(s) $(3x - 1)(x + 4)$
(t) $(3b + 2)(5b + 4)$
(u) $(x - 2y)(x + 2y) - xy$
(v) $(x + 4y)(x - 2y) + 8y^2$
(w) $(a + 1)(b + 1) - ab$
(x) $(n - 1)(m - 1) - nm$
(y) $(2p + 3)(p + 2) - 2p^2$
(z) $(n - 3)(n + 1) + 2n$

3 ABCD is a rectangle.
AB = $(x + 5)$ cm.
BC = $(x + 6)$ cm.
Show that the area, in cm^2 of ABCD is $x^2 + 11x + 30$

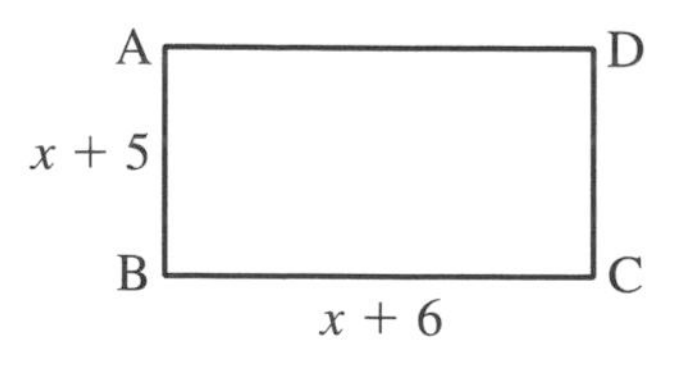

4 PQRS is a rectangle.
PQ = $(x + 3)$ cm.
QR is 5 cm shorter than PQ.
(a) Show that the area, in cm^2, of PQRS is given by the expression $x^2 + x - 6$
(b) Work out the area of PQRS when $x = 8$ cm.

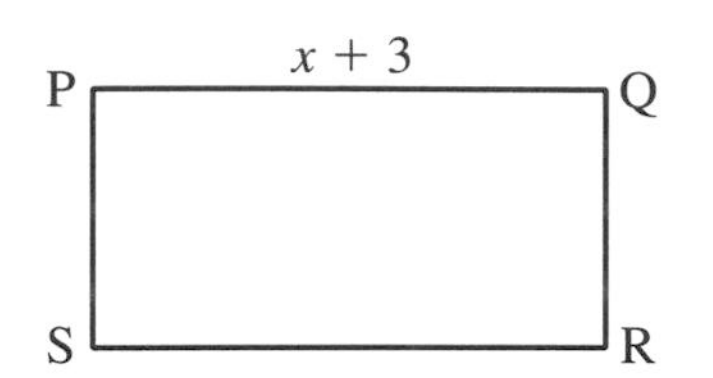

Exercise 21.4 Links: (*21H, I*) 21H, I

1 A formula used in science is

$$v = u + at$$

Work out the value of v when:
(a) $u = 30$, $a = 5$ and $t = 2$
(b) $u = 18$, $a = -2$ and $t = 4$
(c) $u = -12$, $a = -3$ and $t = -7$
(d) $u = -10.5$, $a = -3.8$ and $t = 5$

2 PQRS is a trapezium. The angles at P and Q are both equal to 90°.
The area, A, in cm^2 of PQRS is given by the formula

$$A = \frac{h(a + b)}{2}$$

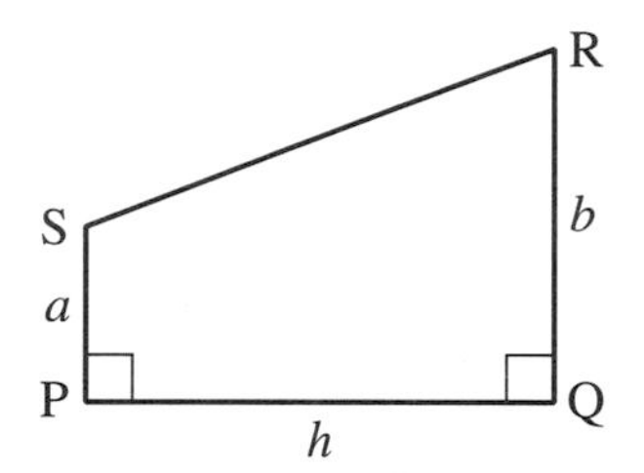

Work out the value of A when:
(a) $a = 3$, $b = 7$ and $h = 8$
(b) $a = 7.6$, $b = 4.8$ and $h = 3$
(c) $a = 8.2$, $b = 5.1$ and $h = 4.6$

3 The formula to calculate the stopping distance, D, in feet for a car travelling at a speed of s in mph is

$$D = 0.05s^2 + s$$

Work out the value of D when s equals:
(a) 20 **(b)** 70 **(c)** 45 **(d)** 37 **(e)** 68

At the scene of an accident, skid marks indicate that a car involved in the accident took 350 feet to come to a stop.
(f) Explain whether or not there is evidence to suggest that the car was breaking the speed limit of 70 mph.

4 The formula to change temperatures measured in degrees Fahrenheit (°F) to degrees Celsius (°C) is

$$C = \frac{5(F - 32)}{9}$$

(a) Work out the value of C when F equals:
(i) 32 **(ii)** 212 **(iii)** 58 **(iv)** 100 **(v)** −22 **(vi)** −16.6

On July 1st last year the temperature in London was recorded as 86°F and in Corfu on the same day the temperature was recorded as 28.7°C

(b) On July 1st last year, where was it hotter, London or Corfu and by how much?

5 A formula used by opticians is

$$f = \frac{uv}{u - v}$$

Work out the value of f when:

(a) $u = 8, v = 3$ **(b)** $u = 6, v = -4$ **(c)** $u = 4.2, v = -7.8$

6 The cost, £C, of hiring a car for a day and driving it x miles is given by the formula

$$C = 50 + 0.2x$$

Re-arrange this formula to make x the subject.

7 Lee works as a painter and decorator. When he decorates a room he charges customers £15 per hour plus the cost of materials.

(a) Work out how much Lee will charge when he works for 10 hours and the cost of the materials is £80.

Val, Lee's partner, designs a formula for the charge, £C, for a job which will take h hours and for which the cost of the materials is £M.

(b) Write down what Val's formula should be.
(c) Re-arrange this formula to make h the subject.

8 Re-arrange each of these formulae to make the letter in brackets the subject:

(a) $y = 3x + 5$ [x] **(b)** $p = 2q - 3$ [q] **(c)** $v = 4u + 7$ [u]
(d) $a = \frac{1}{2}b + 1$ [b] **(e)** $y = \frac{1}{4}x - 5$ [x] **(f)** $s = \frac{3}{4}t + 4$ [t]
(g) $y = ax + b$ [x] **(h)** $v = at + u$ [a] **(i)** $y = a + 3x$ [x]
(j) $y = a + bx$ [x] **(k)** $v = u + at$ [t] **(l)** $r = mp - q$ [p]
(m) $C = \pi d$ [d] **(n)** $D = 2\pi r + a$ [r] **(o)** $y = 2(x + 1)$ [x]
(p) $b = 3(a - 4)$ [a] **(q)** $P = 2(a + b)$ [a] **(r)** $P = 2(x + y)$ [y]
(s) $y = \frac{x + 3}{4}$ [x] **(t)** $y = \frac{x - 5}{2}$ [x] **(u)** $y = \frac{2x - 7}{3}$ [x]
(v) $m = \frac{5n + 3}{7}$ [n] **(w)** $m = \frac{3 - 2n}{4}$ [n] **(x)** $m = \frac{an + b}{c}$ [n]

9 Rearrange each of these formulae to make x the subject:

(a) $y = x^2$ **(b)** $y = x^2 + 1$ **(c)** $y = x^2 - 3$

(d) $y = x^2 + m$ **(e)** $y = x^2 - p$ **(f)** $y = 2x^2$

(g) $y = 5x^2$ **(h)** $y = \frac{1}{2}x^2$ **(i)** $y = ax^2$

(j) $y = bx^2 + 1$ **(k)** $y = cx^2 - 3$ **(l)** $y = ax^2 + b$

(m) $y = \dfrac{x^2}{3}$ **(n)** $y = \dfrac{x^2}{4}$ **(o)** $y = \dfrac{x^2}{a}$

(p) $y = \dfrac{x^2}{3} + 2$ **(q)** $y = \dfrac{x^2}{4} - 5$ **(r)** $y = \dfrac{x^2}{a} - 1$

(s) $y = \dfrac{x^2}{a} - b$ **(t)** $y = \dfrac{ax^2}{c} + b$ **(u)** $y = \dfrac{ax^2 - b}{c}$

10 The formula for the volume, V, of a cylinder with base radius r and height h is

$$V = \pi r^2 h$$

Rearrange this formula to make:

(a) h its subject

(b) r its subject

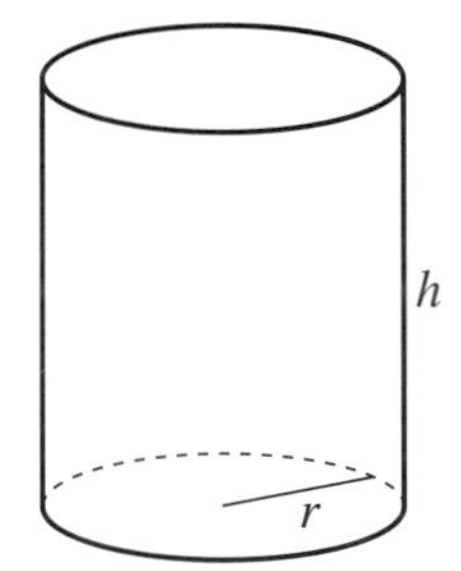

11 **(a)** Show that $\dfrac{1}{3} + \dfrac{1}{5} = \dfrac{5 + 3}{5 \times 3} = \dfrac{8}{15}$

(b) Make v the subject of the formula $\dfrac{1}{f} = \dfrac{1}{v} - \dfrac{1}{u}$

12 Simplify the right-hand side of each formula, then rearrange to make x the subject:

(a) $y = (3x - 2) + 2(x + 5)$

(b) $y = 4(3 - 2x) + 17x$

(c) $y = 3x^2 - 2 + 5 - x^2$

(d) $y = x(3 - x) + x^2$

13 Re-arrange the formula to make t the subject.

$s = at^2 + b$

14 Given that

$$ax + b = cx + d$$

re-arrange to express x in terms of a, b, c and d.

15 $t = 2\pi\sqrt{\dfrac{2}{g}}$.

Re-arrange the formula to make g the subject.

Exercise 21.5 Links: (*21J, K*) 21J, K

1 Factorize each of these expressions:

(a) $3x + 6$ **(b)** $4x - 12$ **(c)** $25p + 5$
(d) $8a + 4$ **(e)** $ab + a$ **(f)** $3x + 12y$
(g) $4a + 12b$ **(h)** $120x + 6$ **(i)** $3x + 6y + 12z$
(j) $3xy + 9x$ **(k)** $10 - 5y$ **(l)** $7a + 14b$
(m) $3 - 9x$ **(n)** $12p - 3q$ **(o)** $250n + 25m$

2 Factorize each of these:

(a) $5x^2 + 4x$ **(b)** $3y - 6y^2$ **(c)** $4b^2 + b$
(d) $3z - z^2$ **(e)** $p^2 - 2p$ **(f)** $n - n^2$
(g) $m^2 + 2m$ **(h)** $x^2 - 3x$ **(i)** $x^3 - 2x$
(j) $a^3 + a$ **(k)** $2x^3 + 3x$ **(l)** $x^2 - 4x$
(m) $4y^4 + 3y$ **(n)** $m^2 - m$ **(o)** $2t + 3t^2$

3 Factorize completely:

(a) $6x^2 + 3x$ **(b)** $4y - 8y^2$ **(c)** $3x - 15x^2$
(d) $7p^2 - 2p$ **(e)** $7b^2 + 3b$ **(f)** $5y - 3y^2$
(g) $4x + 3x^2$ **(h)** $4xy + 12y$ **(i)** $x^3 + 6x^2$
(j) $ab + a^2$ **(k)** $24x^3 - 18x$ **(l)** $ax - a$
(m) $xy + 3x$ **(n)** $x^2y + xy$ **(o)** $a^2x + 3ax$
(p) $6p^2 - 9p$ **(q)** $a^2b + ab$ **(r)** $p^2q + 2pq^2$
(s) $4x^3y + 6xy$ **(t)** $3pq^2 + 2pq$ **(u)** $\pi r^2 + 2\pi rh$
(v) $2x^3y^2 + 4xy$ **(w)** $ut + gt^2$ **(x)** $6a^2b^3 + 15ab^4$

4 Factorize completely:

(a) $8x^2y + 12xy^2$ **(b)** $3ab - 9b$ **(c)** $3x^2y + 6xy^2 - 9xy$
(d) $5a^3b + 15ab$ **(e)** $15xy + 24x^2$ **(f)** $3abc + ab^2c + abc^3$

5 Factorize each of these quadratic expressions:

(a) $x^2 + 8x + 15$ **(b)** $x^2 + 9x + 14$ **(c)** $x^2 + 5x + 4$
(d) $x^2 - 8x + 15$ **(e)** $x^2 - 8x + 12$ **(f)** $x^2 - 7x + 6$
(g) $x^2 + 6x + 9$ **(h)** $x^2 + 8x + 16$ **(i)** $x^2 + 14x + 49$
(j) $x^2 - 12x + 36$ **(k)** $x^2 - 8x + 16$ **(l)** $x^2 - 10x + 25$
(m) $x^2 + 3x - 10$ **(n)** $x^2 - 3x - 10$ **(o)** $x^2 + 3x - 28$
(p) $x^2 + 5x - 14$ **(q)** $x^2 - 3x - 28$ **(r)** $x^2 + x - 30$
(s) $x^2 - x - 2$ **(t)** $x^2 + x - 2$ **(u)** $x^2 - 2x - 3$
(v) $x^2 - 8x - 9$ **(w)** $t^2 + 3t - 40$ **(x)** $y^2 - 9y - 22$

6 Factorize each of these quadratic expressions:

(a) $x^2 - 16$ **(b)** $x^2 - 49$ **(c)** $x^2 - 100$

(d) $x^2 - 1$ **(e)** $5x^2 - 20$ **(f)** $4x^2 - 100$

(g) $2x^2 - 18$ **(h)** $3x^2 - 75$ **(i)** $4x^2 - 36$

7 Factorize completely each of the following:

(a) $x^2 - 7x$ **(b)** $x^2 - 64$ **(c)** $a^2 - 6a + 9$

(d) $x^2 + 12x + 35$ **(e)** $y^2 + 5y - 24$ **(f)** $2n^2 - 8$

(g) $m^2 - m - 30$ **(h)** $2x^2 + 16x + 30$ **(i)** $x^2 + 8x - 20$

(j) $a^3 - a$ **(k)** $x^2 - x - 42$ **(l)** $x^2 - 2x + 1$

Exercise 21.6 Links: (*21L*) 21L

1 Tony repairs cars. He charges:

the cost of the spare parts plus
£15 for every hour that he works on a car.

Tony repairs Mr Akram's car. The cost of the spare parts is £150 and Tony works on this car for h hours.
Tony charges Mr Akram £C.
(a) Write down a formula connecting C and h.
(b) Work out the value of C when $h = 6$. [E]

2 Make y the subject of the formula $x + 2y = 6$. [E]

3 The volume, V, of the barrel is given by the formula:

$$V = \frac{\pi H(2R^2 + r^2)}{3000}$$

$\pi = 3.14$, $H = 60$, $R = 25$ and $r = 20$.
Calculate the value of V. [E]

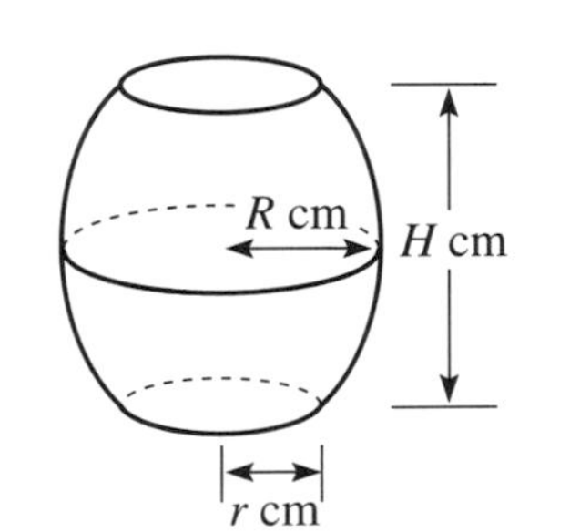

4 **(a)** Factorize completely:
(i) $4st + 8tu - 2tv$
(ii) $15x + 3x^2$
(b) Expand:
$2a(4 - a)$
(c) Expand and simplify:
$(2c + 3)(c - 4)$ [E]

5 $y = ab + c$
Calculate the value of y when $a = \frac{3}{4}$, $b = \frac{7}{8}$ and $c = -\frac{1}{2}$.

Give your answer in the form $\frac{p}{q}$ where p and q are integers. [E]

6 The cost of hiring a boat is:

£80 per day for each day up to and including 7 days and then £60 per day for each day after the first 7 days.

(a) A boat is hired for x days. Write down, in terms of x, expressions for the cost of hiring a boat when
(i) $x \leqslant 7$ **(ii)** $x > 7$

(b) n people hire a boat for x days, x is greater than 7. They share the cost equally to work out the cost per person. The cost per person of hiring the boat is £C. Write down a formula for C in terms of n and x. [E]

7 **(a)** $y = x^2 - 2$. Work out y when $x = -1$.
(b) Factorize $x^2 - 4x - 12$. [E]

22 Percentages

Exercise 22.1 Links: (22*A*, *B*, *C*) 22A, B, C

1 Write these percentages as:
(i) decimals **(ii)** fractions

(a) 40% **(b)** 65% **(c)** 28% **(d)** $66\frac{2}{3}\%$

(e) $62\frac{1}{2}\%$ **(f)** 88% **(g)** $6\frac{1}{3}\%$ **(h)** $37\frac{1}{2}\%$

2 Write these numbers as percentages:

(a) $\frac{1}{4}$ **(b)** 0.8 **(c)** $\frac{13}{40}$ **(d)** $\frac{19}{25}$

(e) 0.41 **(f)** 0.056 **(g)** $\frac{3}{8}$ **(h)** $\frac{2}{5}$

(i) 0.3125 **(j)** 2.69 **(k)** $\frac{17}{20}$ **(l)** $\frac{23}{60}$

3 Write these in order of size, smallest first:

(a) 24%, $\frac{1}{4}$, 0.23 **(b)** 0.60, $\frac{5}{8}$, 62%

(c) $\frac{4}{5}$, 0.79, 81% **(d)** 42%, $\frac{7}{16}$, 0.44

4 52% of the people shopping at a supermarket are women. What percentage are not women?.

5 A hockey team won 38% of their matches and drew 25%. What percentage of their matches did they lose?

6 87% of trains arriving at Shapton are on time. What percentage of trains are not on time?

Exercise 22.2 Links: (22*D*, *E*) 22D, E

1 Work out 20% of £70.

2 Work out 45% of 560 g.

3 Work out $87\frac{1}{2}\%$ of £120.

4 Work out 62% of 600 kg.

5 Work out 7% of £56.

6 Kanti earns £42 500 a year. He pays 6% to the pension fund. How much does he pay to the pension fund?

7 A cinema has 850 seats. 88% of the seats are sold. How many seats are sold?

8 Hersha earns £352 a week. She saves 26%.
How much does she save?

9 Increase £150 by 25%.

10 Increase £32 500 by 4%.

11 Decrease £95.20 by 10%.

12 Decrease £840 by 60%.

13 Peter's council tax last year was £840. The council tax has risen by 6%. What is the council tax this year?

14 What is the sale price of this dress normally costing £150?

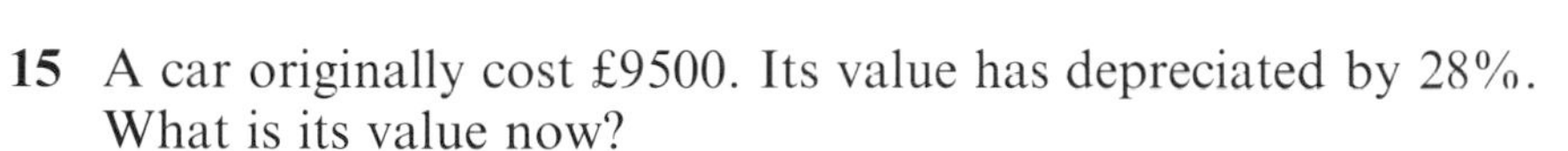

15 A car originally cost £9500. Its value has depreciated by 28%.
What is its value now?

Exercise 22.3 Links: (*22F, G*) 22F, G

1 Rodger hits 28 clays out of 40.
Write this as a percentage.

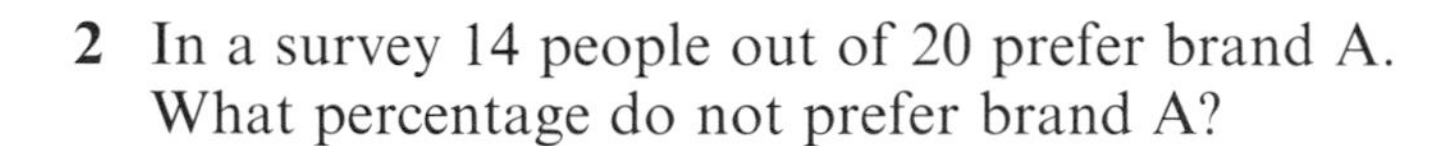

2 In a survey 14 people out of 20 prefer brand A.
What percentage do not prefer brand A?

3 Express the first quantity as a percentage of the second:
(a) 40 kg, 120 kg **(b)** 32 cm, 4 m **(c)** £1.35, £4.00

4 The cost of a season ticket has risen from £252.00 to £263.34.
Work out the increase as a percentage of the original cost.

5 The cost of a hi-fi system has fallen from £499 to £449. What is the percentage decrease in price?

6 Margaret buys a dress in a sale. It has been reduced by 20% to £70.40 in a sale. What was its original price?

7 Amelia's weekly wage has increased by 3.5% to £74.52. What weekly wage did she receive before the increase?

8 Andrew's pet dog weighs 7.5 kg. It used to weigh 6 kg. What is the percentage increase?

Exercise 22.4 Links: (*22H, I, J*) 22H, I, J

1 For each of these find the percentage profit:
(a) cost price = £30, selling price = £45
(b) cost price = £50, selling price = £120

2 For each of these find the percentage loss:
(a) cost price = £40, selling price = £30
(b) cost price = £65, selling price = £48

3 A shop buys a toaster for £8 and sells it for £14. Find the percentage profit.

4 Christine buys a car for £4800 and sells it a year later for £3000. Find her percentage loss.

5 Work out the VAT at $17\frac{1}{2}\%$ on these prices:
(a) £5 **(b)** £22 **(c)** £99.50
(d) £237.50 **(e)** £45.60 **(f)** £64.90

6 A car radio is advertised as £85 + VAT. Work out the total cost including VAT.

7 These prices include VAT.
Work out the price exclusive of VAT:
(a) £141 **(b)** £9.40 **(c)** £423

8 VAT is added to the price of a £42 food mixer. Find the total cost.

9 Find **(i)** the total cost of buying on credit
(ii) the difference between the cash price and the cost of buying on credit

(a) A secondhand car costs £3450. It can also be bought with a deposit of 20%, followed by 24 monthly payments of £138.
(b) A cooker costs £499. It can also be bought on credit with 12 monthly payments of £39.50 and a deposit of 15%.
(c) A new patio for a garden costs £1020. It can also be bought for a deposit of 35% and 36 monthly payments of £34.50.

Exercise 22.5 Links: (*22K, L, M*) 22K, L, M

1 Find the simple interest when:
(a) £1000 is invested for 5 years at 6% p.a.
(b) £565 is invested for 3 years at 8% p.a.
(c) £4790 is invested for 6 years at 11% p.a.
(d) £7500 is invested for $5\frac{1}{2}$ years at 5.5% p.a.

2 Find the time it takes for a sum of £500 to produce £60 simple interest at $6\frac{1}{4}$% p.a.

3 Find the sum of money that should be invested to earn £500 simple interest over 2 years at $7\frac{1}{4}$%.

4 Work out the time it takes for £140 to produce £50 simple interest at 9% p.a.

5 £800 is invested for 3 years at 5% compound interest which is paid annually. What is the total interest earned?

6 £5600 is invested for 2 years at $8\frac{1}{2}$% compound interest which is paid annually. What is the total interest earned?

7 A new caravan is £10 500 now. Each year the price decreases by 4% of the price at the beginning of the year. Calculate the price of the caravan in 4 years' time.

8 £380 is invested for two years at $9\frac{1}{2}$% p.a. compound interest which is paid every six months. What is the total interest earned?

9 Repeat questions **5** and **6** using the formula

$$A = P\left(1 + \frac{R}{100}\right)^n$$

Exercise 22.6 Links: (22*N*, *O*) 22N, O

1 Russell is paid a basic rate of £3.20 per hour, and a $12\frac{1}{2}$% commission on sales.
How much is he paid if he works 35 hours and makes sales of £450 during one week?

2 Interest on a loan is charged at 12.5% p.a. How much interest is paid each month on a loan of £2600 if there are 12 equal monthly payments during the year?

3 Garden chairs are priced at £12 + $17\frac{1}{2}$% VAT each, with a discount of 20% of the total price before VAT if six or more chairs are bought. How much would it cost to buy 8 chairs?

4 The price of a computer is decreased from £1450 by 15%. What is the new price?

5 What is the percentage saving on this sofa?

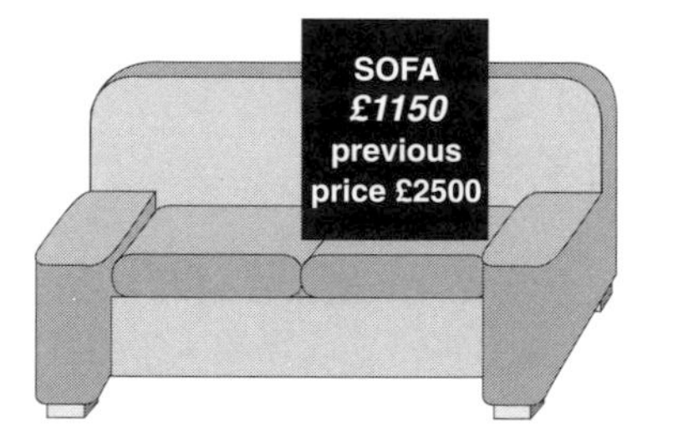

6 Cathy buys a new kitchen. She pays £250 deposit and 36 monthly payments of £84.50.
(a) Work out the total amount Cathy pays.

In Blackstone's kitchen shop a kitchen is priced at £2800. Neil pays cash and is given a discount of 8%.
(b) Calculate the amount that Neil pays.

The price of the Blackstone's kitchen rises the following year by 4.5%.
(c) Work out the new price.

7 Clancy buys a new gas cooker on credit.
He pays 25% deposit and 24 monthly payments of £15.75.
Work out
(a) the total cost of buying the cooker on credit
(b) the amount Clancy would save if he bought the cooker with cash.

8 In a sale a coat originally costing £180 is reduced by 25%. It still does not sell and the shop reduces it by a further 20%. What is the new sale price?

23 Transformations

Exercise 23.1 Links: (*23A–G*) 23A–G

1 Copy this grid into your exercise book.
Translate the shape T using the following vectors.

(a) $\begin{pmatrix} 5 \\ 2 \end{pmatrix}$, call it A **(b)** $\begin{pmatrix} 5 \\ -2 \end{pmatrix}$, call it B.

(c) $\begin{pmatrix} -4 \\ -2 \end{pmatrix}$, call it C **(d)** $\begin{pmatrix} -1 \\ 1 \end{pmatrix}$, call it D.

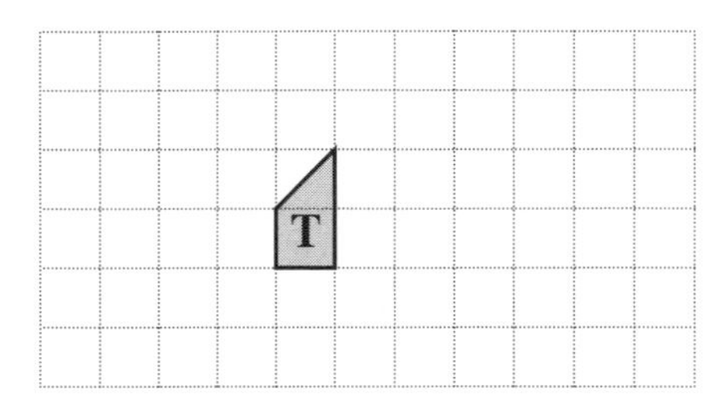

2 The shape S has been moved into four different positions A, B, C, D.
Write down the translation vectors for each of the four translations.

3 Copy this diagram into your exercise book.
For each shape reflect it in the mirror line **m**.

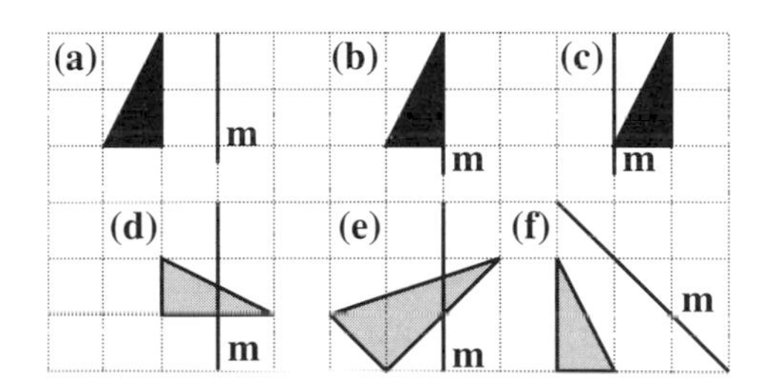

4 Shape S has been reflected four times.
Describe fully each of these reflections that takes S to positions A, B, C, D.

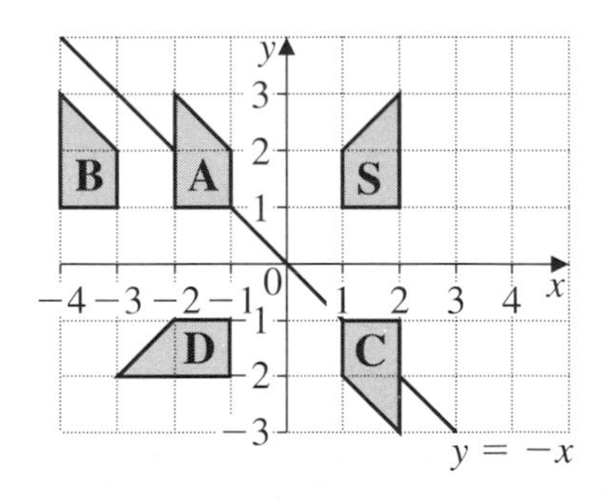

5 Copy this diagram into your exercise book.
Rotate each triangle about the point P marked with a cross, by the angle and direction given.
(a) 90° clockwise
(b) 270° anticlockwise
(c) 180° anticlockwise
(d) 90° anticlockwise
(e) 180° clockwise
(f) 45° clockwise

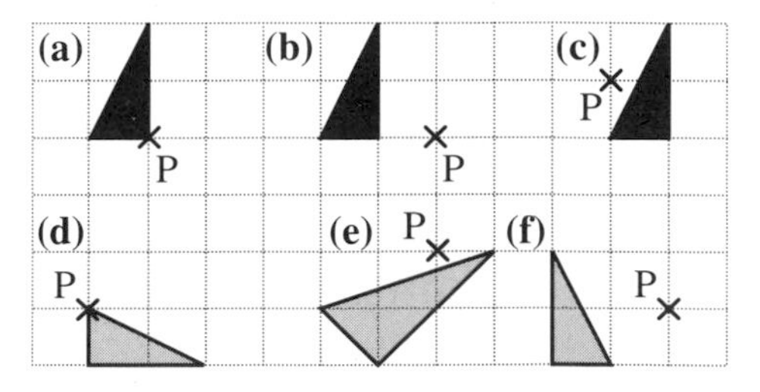

6 Shape S has been rotated four times.
Describe fully each of these rotations that takes S to positions A, B, C, D.

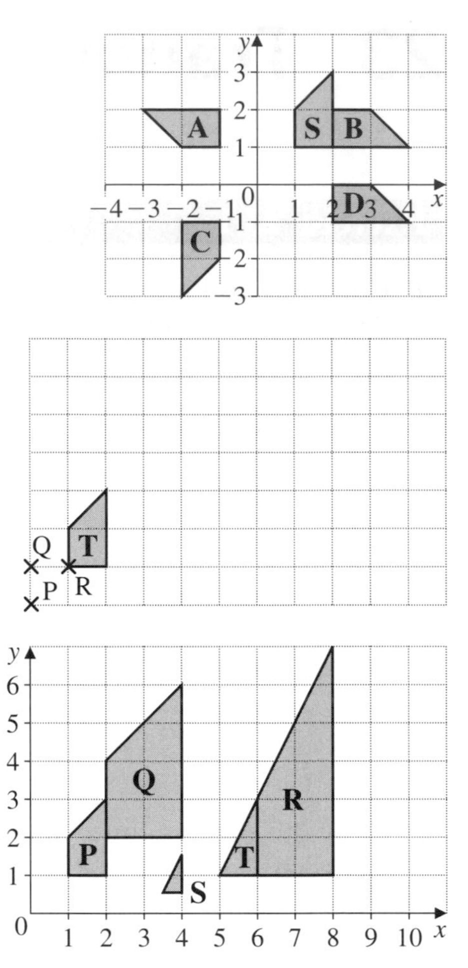

7 Copy this diagram into your exercise book.
Enlarge shape T by the following:
(a) scale factor 2 from point P
(b) scale factor 3 from point Q
(c) scale factor 2.5 from point R
(d) scale factor $\frac{1}{2}$ from point R.

8 Describe the enlargement that moves:
(a) shape P on to shape Q
(b) shape T on to shape R
(c) shape T on to shape S
(d) shape R on to shape T

9 With P as the centre of enlargement, draw the shaded shape after an enlargement by scale factor $\frac{2}{3}$.

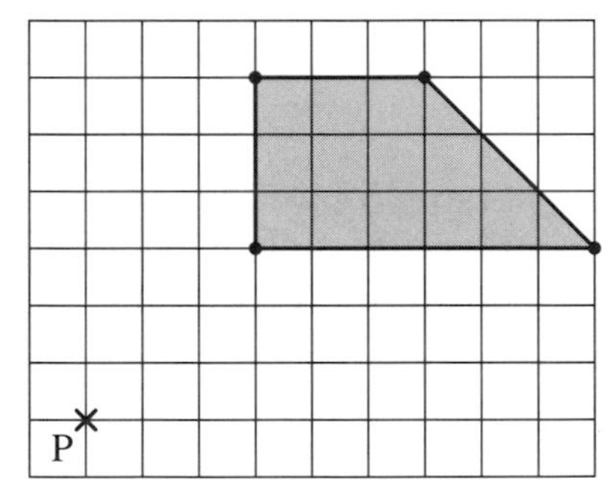

Exercise 23.2 Links: (*23H*) 23H

1 **(a)** Triangle T is transformed on to triangle R by a single transformation. Write down in full this single transformation.
(b) Triangle R is transformed on to triangle M by a single transformation. Write down in full this single transformation.
(c) Write down in full the single transformation that will move triangle M back on to triangle T.

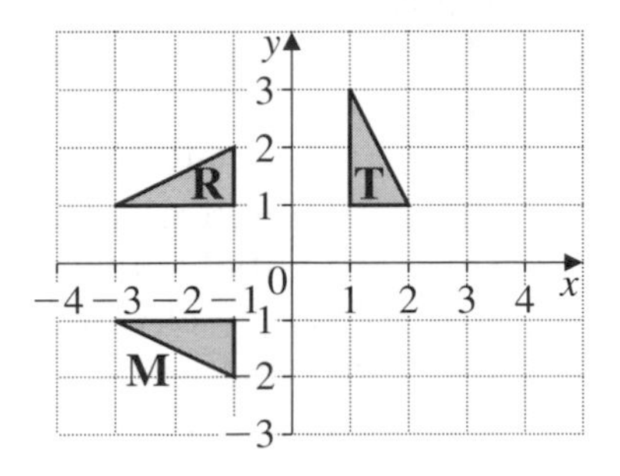

2 On a coordinate grid with both x and y axes drawn from -4 to $+4$ plot the points (1, 1), (1, 3), (2, 2) and (2, 1). Label this shape A.

(a) Transform the shape A by reflecting it in the x-axis. Label the shape B.

(b) Rotate shape B by rotating it 90° clockwise centre (0, 0). Label this shape C.

(c) Write down the single transformation that will move shape C back on to shape A.

3 On a coordinate grid with both x and y axes drawn from -4 to $+4$ plot the points (1, 1), (1, 3), (2, 2) and (2, 1). Label this shape A.

(a) Transform the shape A by reflecting it in the y-axis. Label the shape B.

(b) Rotate shape B by rotating it 90° clockwise centre (0, 0). Label this shape C.

(c) Write down the single transformation that will move shape C back on to shape A.

4 On a coordinate grid with both x and y axes drawn from -6 to $+6$ plot the points (1, 1), (1, 3), (2, 2) and (2, 1). Label this shape A.

(a) Transform the shape A by enlarging it scale factor 2 centre (0, 0). Label the shape B.

(b) Rotate shape B by rotating it 90° clockwise centre (0, 0). Label this shape C.

(c) Reflect shape C by reflecting it in the y-axis. Label this shape D.

(d) Write down the single transformation that will move shape D back on to shape B.

5 On a coordinate grid with both x and y axes drawn from -4 to $+4$ plot the points (1, 1), (1, 3), (2, 2) and (2, 1). Label this shape A.

(a) Rotate shape A by rotating it 90° anticlockwise centre (0, 0). Label this shape B.

(b) Transform the shape B by reflecting it in the x-axis. Label the shape C.

(c) Reflect shape C in the line $y = x$. Label the reflected shape D.

(d) Transform shape D by rotating it 180° centre $(-1, -1)$. Label the new shape E.

(e) Write down the single transformation that will move shape E back on to shape A.

6 Copy the diagram.

(a) Reflect the triangle T in the line $y = x$. Label the shape R.

(b) Rotate the triangle T centre (0, 0) through 180°. Label the shape S.

(c) Describe fully the transformation that will move shape S on to shape R.

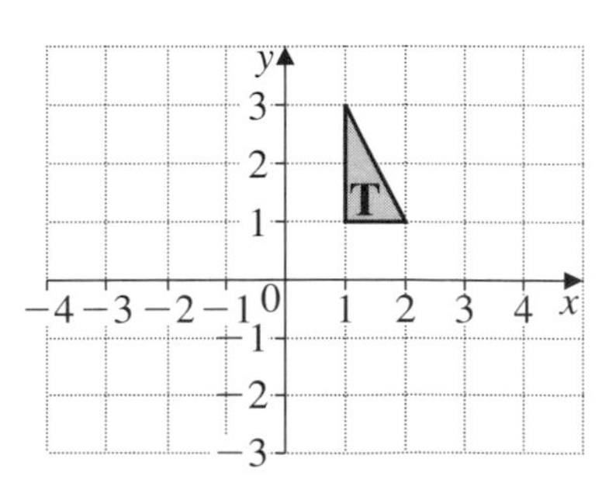

24 Presenting data

Exercise 24.1 Links: (*24A*, *B*) 24A, B

1 A shop rents out videos. The table shows the number rented out over a 2-month period.

	Week 1	Week 2	Week 3	Week 4	Week 5	Week 6	Week 7	Week 8
Number rented	200	250	300	270	300	425	450	425

(a) Draw a pictogram with each symbol representing 50 videos.

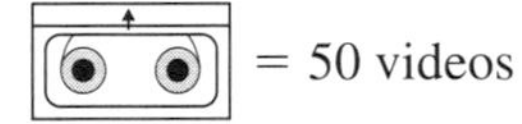

(b) Draw a bar chart and use this to construct a frequency polygon.

2 Newspaper sales are shown in the pictogram.

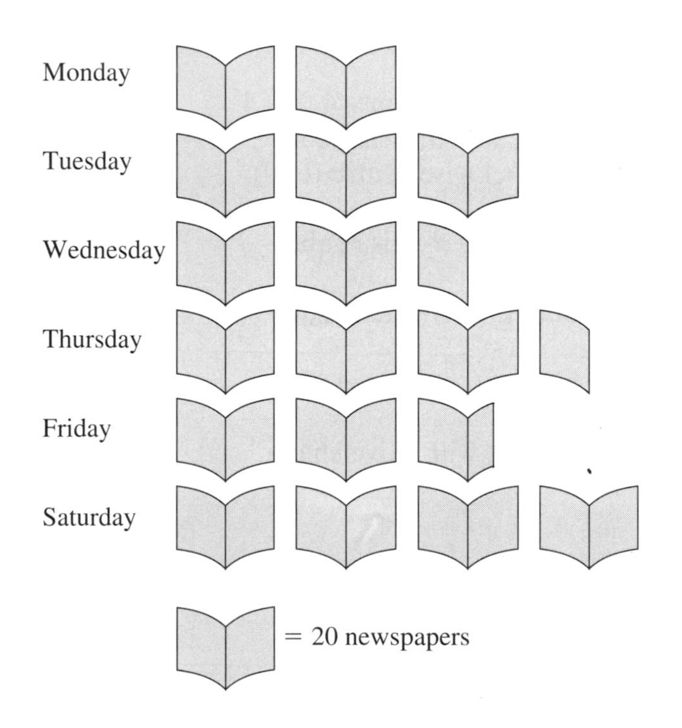

(a) Draw a table and record the frequencies.
(b) Use this to construct a frequency diagram.

3 The table shows the value of the FTSE share price index during 2001.

JULY				AUGUST				SEPTEMBER				
WEEK				WEEK				WEEK				
1	2	3	4	1	2	3	4	1	2	3	4	5
5700	5400	5500	5400	5270	5550	5400	5350	5350	5390	4800	4420	4930

Construct a time series graph from this information.
Comment on the trend.

4 The table shows the frequency distribution of student absences for a year.

Absences d (days)	Frequency
$0 \leqslant d < 5$	4
$5 \leqslant d < 10$	6
$10 \leqslant d < 15$	8
$15 \leqslant d < 20$	5
$20 \leqslant d < 25$	4
$25 \leqslant d < 30$	3

Draw a frequency polygon for this frequency distribution.

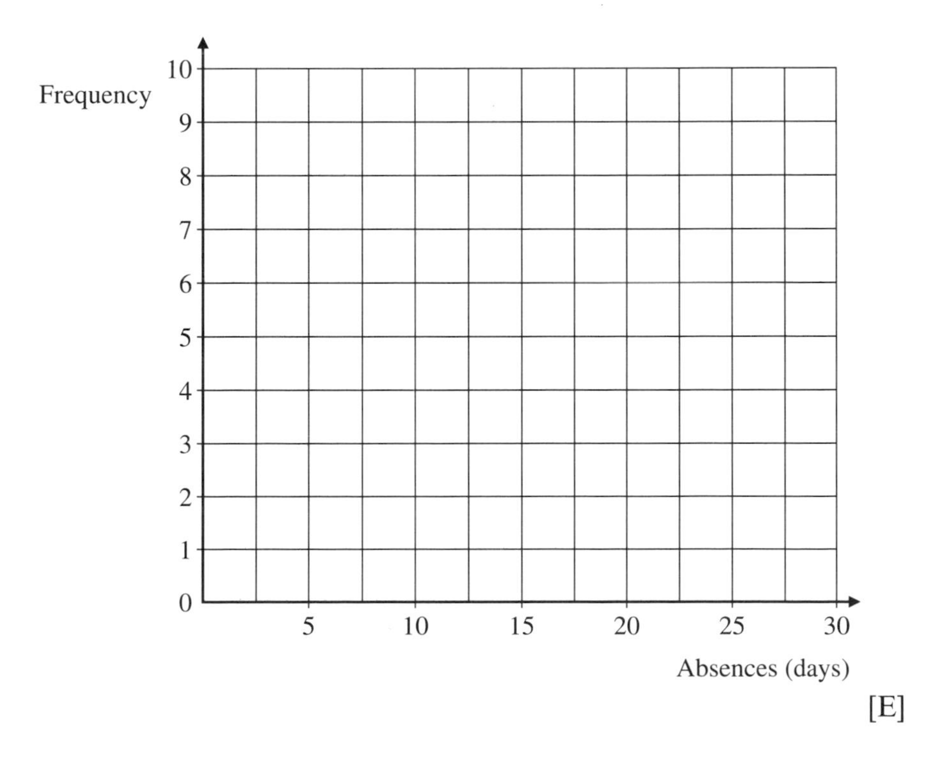

[E]

Exercise 24.2 Links: (*24C*) 24C

1 The table shows how many different animals are on Ben's farm.

Animal	Number
Cows	36
Pigs	24
Sheep	75
Chickens	45

Animal	Angle
Cows	
Pigs	
Sheep	
Chickens	

(a) Work out the angle of each sector of the pie chart.
(b) Construct the pie chart. [E]

2 Patrick carried out a survey of 45 pupils in Year 11. He asked them how many books they had borrowed from the library in the last month. The maximum number anyone can borrow is 12. The frequency table shows his results.

Number of books	Frequency
0 to 2	11
3 to 5	15
6 to 8	13
more than 8	6

(a) Construct a pie chart to show this information.
(b) What fraction of the group had borrowed 6, 7 or 8 books?
(c) What percentage had borrowed 6 or more books?
(d) Calculate an estimate of the mean number of books borrowed. [E]

3 The pie charts shows how Jenny spends her monthly income.

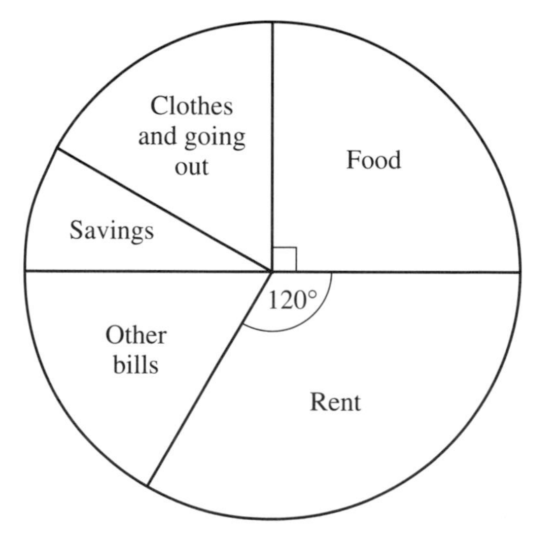

Jenny spends £150 a month on food.
(a) Work out Jenny's monthly income.
(b) Work out how much rent Jenny pays each month. [E]

Exercise 24.3 Links: (*24D*) 24D

1 The table below contains information about first class rail fares and distances of towns and cities from London.

Town/city	Distance in km	First class fare in £
Ashford	56	21.90
Bangor	239	69.50
Bristol	118	51.00
Cardiff	145	58.00
Carlisle	299	90.00
Derby	128	45.50
Doncaster	156	71.00
Felixstowe	84	27.50
Harrogate	204	84.50
Hull	197	73.00
Kettering	72	25.50
Lincoln	136	51.50
Leicester	99	37.00
Manchester	183	70.00
Liverpool	193	68.50
Norwich	115	43.50
Plymouth	226	72.00
York	188	83.50

(a) Draw a scatter graph to show this information.
(b) Describe the correlation.
(c) Draw a line of best fit.
(d) Estimate the first class fare to Chester (179 km from London).
(e) The first class fare to Exeter from London is £64. Estimate how far Exeter is from London.

2 The table shows details of attendance and entrance charges for visitors to the properties of a national organisation.

Attendance ('000s)	Charge (£)	Attendance ('000s)	Charge (£)
297	4.20	211	6.00
200	6.00	188	3.00
180	3.50	158	5.00
141	5.50	114	5.00
108	4.90	96	2.30
81	3.60	82	5.40
67	4.00	50	3.30

(a) Draw a scatter diagram and describe the correlation.
(b) Give possible reasons for your answer.

3 Joe has 12 cars for sale. The scatter diagram shows the ages and prices of the cars.

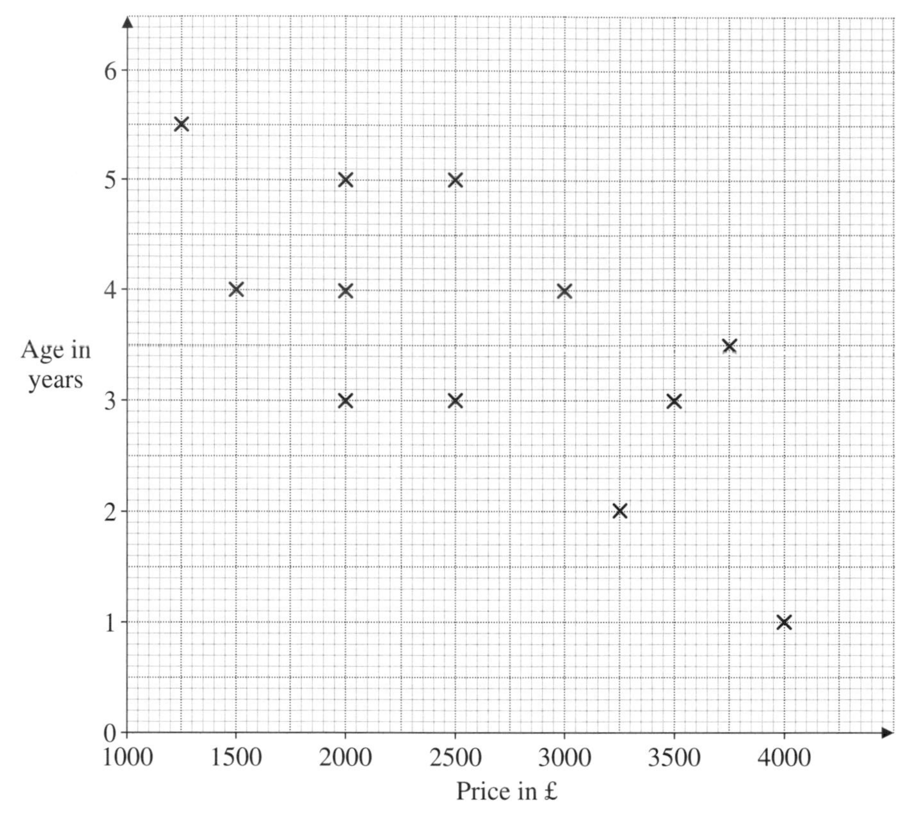

Joe buys five more cars to sell.
The table lists the ages and prices of these cars.

Age in years	3	4	1	5	4
Price in £	3000	2500	3750	1500	2250

(a) Plot this information on the scatter graph.
(b) Describe the correlation between the ages of these seventeen cars and their prices. [E]

Exercise 24.4 Links: (*24E*) 24E

1 Janet receives a weekly bank statement to allow her to keep track of her money. The table shows the balances for the first 3 months.

Week	1	2	3	4	5	6	7
	223	191	1306	728	315	280	1407
Week	8	9	10	11	12	13	
	834	420	348	1492	910	452	

Draw the time series as a graph and impose a suitable moving average on it. Comment on the trends and make some speculation about Janet's income and expenditure.

2 The table gives information about passenger numbers on an internal airline route one week.

	Flight							
	Morning		Midday		Afternoon		Evening	
	Bus. Class	Economy	Bus. Class	Economy	Bus. Class	Economy	Bus. Class	Economy
Monday	10	58	2	103	5	74	14	63
Tuesday	8	63	3	89	4	75	17	58
Wednesday	9	72	3	95	5	80	15	71
Thursday	7	60	4	90	3	69	12	49
Friday	2	43	1	72	9	80	20	72

Create four-point moving averages for each of the categories business class and economy.

25 Ratio and proportion

Exercise 25.1 — Links: (*25A*, *B*) 25A, B

1 Write these ratios in their simplest form:

(a) 30 : 20 **(b)** 18 : 6 **(c)** 24 : 15
(d) 10 : 4 **(e)** 36 : 24 **(f)** 8 : 20
(g) 12 : 14 **(h)** 9 : 48 **(i)** 143 : 1001
(j) 5 : 10 : 15 **(k)** 48 : 24 : 6 **(l)** 14 : 56 : 35

2 Write these ratios in their simplest form:

(a) 2 cm : 1 m **(b)** 450 mg : 1 g **(c)** 25 m*l* : 2 *l*
(d) 2 kg : 50 g **(e)** 48 p : £2 **(f)** 40 secs : 1 min
(g) 55 cm : 2 m **(h)** 12 mm : 1 m **(i)** £2 : 40 p : £3.60
(j) 1 mm : 1 cm : 1 m **(k)** 82 p : £5 **(l)** 20 min : 2 h : $\frac{1}{4}$ h

3 Write these ratios in their simplest form:

(a) $3 : \frac{1}{4}$ **(b)** $4 : \frac{1}{2}$ **(c)** $\frac{3}{4} : \frac{1}{2}$
(d) $1\frac{1}{2} : 2$ **(e)** $\frac{1}{3} : \frac{1}{2}$ **(f)** $\frac{3}{4} : 2\frac{1}{4}$
(g) $\frac{3}{4} : \frac{5}{8} : \frac{1}{2}$ **(h)** $\frac{3}{5} : \frac{7}{8}$ **(i)** $\frac{3}{4} : 3\frac{1}{2} : 2$

4 Write each of these ratios in their simplest form:

(a) 1.5 : 4 **(b)** 2.6 : 3.8 **(c)** 0.06 : 2
(d) 4.2 : 2.4 **(e)** 0.01 : 2.5 **(f)** 0.06 : 0.3 : 1
(g) 6.8 : 0.004 **(h)** 2 : 2.5 : 3.5 **(i)** 5 : 0.6 : 0.12

5 Write each of these ratios in the form $1 : n$

(a) 2 : 7 **(b)** 3 : 15 **(c)** 6 : 15
(d) 5 : 13 **(e)** 20 cm : 2 m **(f)** 40 g : 1 kg
(g) 45 p : £2 **(h)** 56 p : £3 **(i)** 30 min : 2 h

6 Write each of these ratios in the form $n : 1$

(a) 20 : 4 **(b)** 42 : 5 **(c)** 6 : 12
(d) 6 : 15 **(e)** 8 : 6 **(f)** 4 h : $\frac{1}{2}$ h
(g) 2 *l* : 10 m*l* **(h)** 4 m : 2 cm **(i)** 50 p : £2

Exercise 25.2 — Links: (*25C*, *D*) 25C, D

1 Find x for each of these pairs of equivalent ratios:

(a) $x : 4$ 16 : 24 **(b)** 4 : 13 $20 : x$ **(c)** $x : 6$ 14 : 42
(d) 18 : 6 $x : 1$ **(e)** 9 : 45 $1 : x$ **(f)** $9 : x$ 45 : 80
(g) $\frac{1}{2} : x$ 10 : 20 **(h)** 3 : 8 $x : 5$ **(i)** $5 : x$ 35 : 63

2 The ratios 3 : 5 : x and 24 : 40 : 56 are equivalent. Find x.

3 The ratio of girls to boys in a drama group is 7 : 6. There are 24 boys in the drama group.
How many girls are there in the drama group?

4 The ratio of the length to the width of a rectangle is 5 : 4.
The width of the rectangle is 12 cm.
Work out the length of the rectangle.

5 A drink contains orange juice and water in the ratio 1 : 4.
There are 340 m*l* of water in the drink.
Work out the amount of orange juice in the drink.

6 A concrete mix is made by adding sand and cement in the ratio 5 : 1. Six buckets of cement are put into a mixer. Work out the number of buckets of sand that should be put into the mixer.

7 Divide the quantities in the ratios given:
(a) £16 in the ratio 3 : 5
(b) £24.80 in the ratio 2 : 3
(c) £130 in the ratio 10 : 3
(d) £726 in the ratio 1 : 2 : 3
(e) 3 m in the ratio 3 : 7
(f) 45 cm in the ratio 4 : 5
(g) 96 kg in the ratio 5 : 3
(h) £360 in the ratio 7 : 11
(i) 61.56 m in the ratio 5 : 4
(j) £721 in the ratio 1 : 4 : 2

8 The ratio of girls to boys in a class is 7 : 6. There are 26 students in the class.
Find out how many are **(a)** girls **(b)** boys.

9 The ratio of male to female members at Lucea Golf Club is 7 : 5.
The golf club has 900 members.
How many members will be:
(a) male
(b) female?

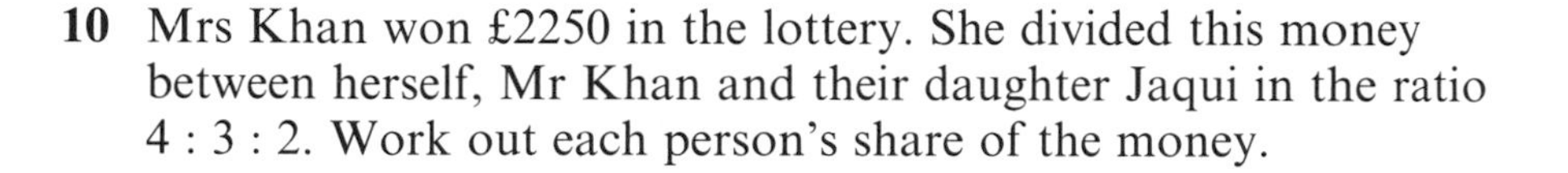

10 Mrs Khan won £2250 in the lottery. She divided this money between herself, Mr Khan and their daughter Jaqui in the ratio 4 : 3 : 2. Work out each person's share of the money.

11 Jan travels 154 miles to visit her aunt. She makes the journey by a mixture of car, rail and bus in the ratio 1 : 4 : 2. How far does she travel:
(a) by car **(b)** by rail **(c)** by bus?

12 Ian divides a length of pipe into three sections which are in the ratio 4 : 5 : 6. The length of the pipe is 3 metres. Work out the length of each of the three sections.

Exercise 25.3 Links: (*25E, F*) 25E, F

1 Zorba bought 12 litres of petrol for his motorbike. The total cost of the petrol was £9.06.
(a) What was the cost of 1 litre of petrol?

Eric bought 16 litres of the same petrol for his car.
(b) How much did Eric pay for this petrol?

2 Barbara is paid £18 for four hours work.
(a) How much is she paid for each hour?
(b) How much should she be paid for
(i) six hours work **(ii)** fifteen hours work?

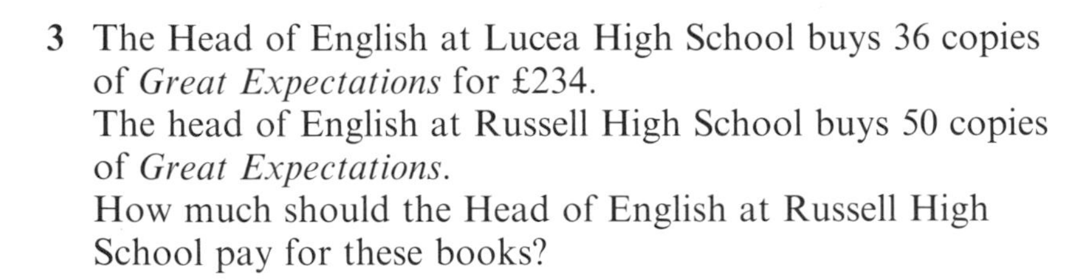

3 The Head of English at Lucea High School buys 36 copies of *Great Expectations* for £234.
The head of English at Russell High School buys 50 copies of *Great Expectations*.
How much should the Head of English at Russell High School pay for these books?

4 Six standard size jars of coffee cost £13.20. Work out the cost of eight standard size jars of the same coffee.

5 Five identical tubes of toothpaste have a total capacity of 600 m*l*. Work out the total capacity of nine of these tubes of toothpaste.

6 When it is full, a trough provides enough drinking water to last 12 horses for 4 days. How many days will the same trough, when full, last for:
(a) 9 horses
(b) 18 horses
(c) 24 horses

7 Six identical machines can complete a job in 5 hours. How long will it take:
(a) one machine to complete the job
(b) twelve machines to complete the job
(c) four machines to complete the job?

8 Using three identical copiers, a printer can make a copy of a manuscript in 20 minutes.
How long will it take the printer to make a copy of the same manuscript if she has:
(a) only two identical copiers **(b)** five copiers

9 A quantity of food will last six men for nine days. How long will the same quantity of food last:
(a) 4 men **(b)** 18 men **(c)** 5 men?

Exercise 25.4 Links: (*25G*) 25G

1 Fatima has 35 CDs and tapes. The ratio of the number of CDs to the number of tapes is 3 : 4.
Work out how many CDs she has. [E]

2 Mr Akram won £5000 in the lottery. He kept 40% of the money for himself. He then divided the rest of the money between his three children in the ratio of their ages. The three children are aged 3, 6 and 15 years.
Work out each child's share of the money.

3 When ten people hire a boat for a week the cost per person is £56. Work out the cost per person when the same boat is hired for a week by:
(a) 5 people **(b)** 7 people **(c)** 8 people

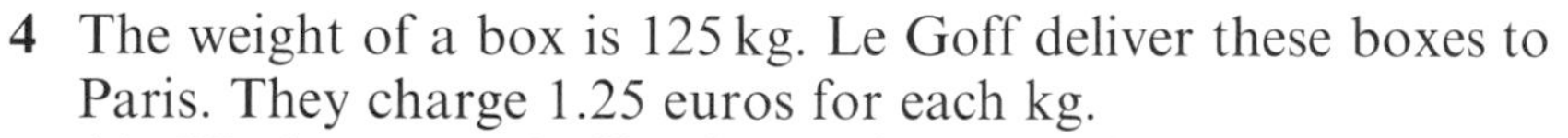

4 The weight of a box is 125 kg. Le Goff deliver these boxes to Paris. They charge 1.25 euros for each kg.
(a) Work out Le Goff's charge, in euros, for delivering one of these boxes.

Le Goff deliver 50 of these boxes to Paris.
€1 = 6.55957 French francs.
(b) Work out Le Goff's charge, in €, for delivering the 50 boxes. Give your answer correct to the nearest €. [E]

5 Given that 5 miles is equivalent to 8 kilometres:
(a) work out, in kilometres
(i) 40 miles **(ii)** 12 miles **(iii)** 240 miles
(b) work out, in miles
(i) 24 kilometres **(ii)** 50 kilometres **(iii)** 3 kilometres

6 The depths of two wells are in the ratio 7 : 9. The depth of the deeper of the two wells is 63 metres. Work out the depth of the other well.

7 The recipe for making 15 cakes is:

Flour 250 g	Butter 175 g	Sugar 175 g
Fruit 200 g	Eggs 3	Milk 330 m*l*

Jenny wants to make 10 of these cakes. Change the amounts given in the recipe to those needed for 10 cakes. Give all your answers to the nearest whole number of units.

8 A 70 c*l* bottle of whisky costs £8.50. A one-litre bottle of whisky costs £11.80. Explain whether or not the costs and capacities are in proportion.

9 The width and length of a rectangular sheet of paper are in the ratio 3 : 5. The width of the sheet of paper is 21 cm.
Work out
(a) the length of the sheet of paper
(b) the area of the sheet of paper

10 Naomi's take home pay is £154 each week.
She divides this money between savings and spending in the ratio 2 : 5.
Work out how much she saves each week.

11 Mr Khan won £4800 in a lottery.
He divided this money between himself, his wife and his daughter in the ratio 4 : 3 : 1.
Work out the amount of money he gave to his wife.

12 Given that 5 litres is approximately 1.1 gallons, work out
(a) the number of litres equivalent to 5.5 gallons
(b) the number of gallons equivalent to 65 litres.

The cost of 1 litre of petrol is 72 p.
(c) Work out the cost of 1 gallon of petrol.

13 John works as a sales representative.
Each week he is paid a wage plus expenses.
Last week his expenses and wage were in the ratio 2 : 9.
John's wage last week was £360.
Work out
(a) John's expenses last week
(b) the total amount he was paid last week.

14 Jennifer paid £8.64 for 12 litres of petrol.
How much should she pay for
(a) 8 litres of petrol
(b) 20 litres of petrol
(c) 55 litres of petrol

15 Divide £450 in the ratio 4 : 3 : 2

26 Accurate drawings, scales and loci

Exercise 26.1 Links: (*26A, B*) 26A, B

1 (a) Make accurate full size drawings of the triangles:

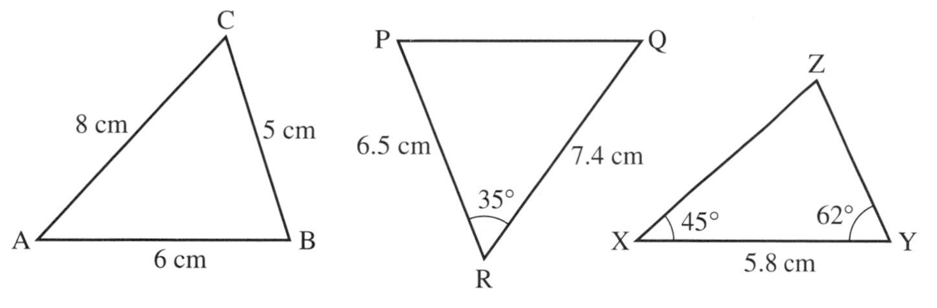

(b) Measure and write down the sizes of all the unmarked lines and angles.

2 Draw accurate full size drawings of these triangles where possible:

(a) DEF with DE = 4.5 cm, EF = 5.4 cm and DF = 10 cm
(b) JKL with JK = 4.5 cm, KL = 5.4 cm and JL = 7.2 cm
(c) MNO with MN = 5.7 cm, MO = 7.3 cm and angle M = 42°
(d) PQR with PQ = 7.5 cm, QR = 6.3 cm and angle P = 33°
(e) TUV with TU = 5.7 cm, angle T = 57° and angle U = 39°
(f) WXY with WX = 7.5 cm, angle W = 60° and XY = 9.2 cm

3 (a) Make an accurate full size drawing of the quadrilateral ABCD.
(b) Measure and write down the size of:
(i) AD
(ii) angles A and B

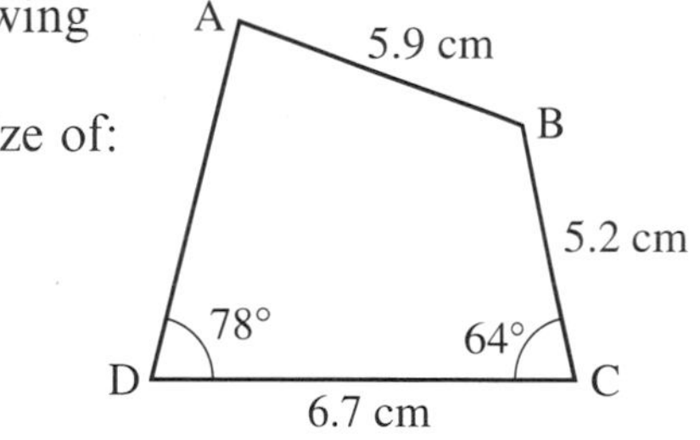

4 Albert makes a model of a fire engine. The scale of the model is 1 : 20. The height of the real fire engine is 4.2 metres.
(a) Calculate the height of the model fire engine.

The length of the model fire engine is 54 cm.
(b) Calculate the length of the real fire engine.

The ladder makes an angle of 15° with the top of the model fire engine.
(c) Work out the angle the ladder makes with the top of the real fire engine.

5 On the Pathfinder map she was using Gill measured the distance she still had to walk as 14.5 cm. The map was drawn on a scale of 1 : 25 000. Calculate the actual distance Gill still had to walk.

6 Caesar was driving along Ermin Street and measured the distance he travelled in a straight line as 12 miles. What distance would that be on a map with a scale of 1 : 50 000.

7 Sue submitted plans for an extension to her house.
The scale of the plans was 1 : 50.
The length of the extension on the plan was 5.3 cm.
(a) What was the actual length of the extension?

The width of the actual extension is 3.1 metres.
(b) What measurement would the width be on the plan?

8 Marlborough is 16 miles from Swindon on a bearing of 170°. Chippenham is 20 miles from Marlborough on a bearing of 280°. Using a scale of 1 cm represents 2 miles make a scale drawing of these places.
(a) Find the bearing of Swindon from Chippenham.
(b) Calculate the distance of Swindon from Chippenham.

Exercise 26.2 Links: (*26C*) 26C

1 Construct the locus of the following points:
(a) 3 cm from the point Q
(b) equidistant from X and Y where XY = 5 cm
(c) equidistant from the lines AB and BC where angle ABC = 60°
(d) 2 cm from the straight line PQ where PQ = 7.5 cm.

2 A goat is tethered by a 10-metre-long chain in the middle of a large field. Draw, using a scale of 1 cm to represent 4 metres the locus of the area that the goat can graze in if the chain is attached:
(a) to a tree **(b)** to a bar that is 20 metres long.

3 Ermintrude the cow is attached by a 15-metre-long chain to a bar that runs along the whole length of the long side of a barn that is located in the middle of a large field, as shown in the diagram. Using a scale of 1 cm to represent 2 metres draw the locus of the area in which she can graze.

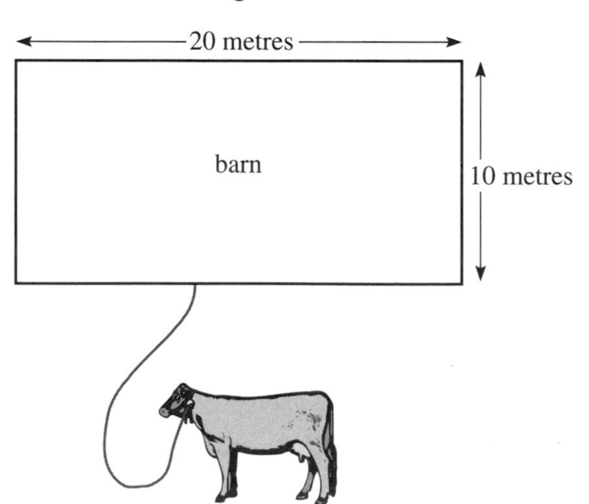

4 Pegleg, the mathematical pirate, buried his treasure according to the following instructions.

> 'The treasure lies on a bearing of 150° from the tall coconut tree and a distance of 50 metres from the spiky cactus plant.'

The spiky cactus plant is 20 metres due East of the coconut tree.

(a) Draw a scale drawing to show the position of the treasure using a scale of 1 cm to represent 5 metres.

Captain Pugwash finds the treasure and re-buries it at a point that is equidistant from the tall coconut tree and the spiky cactus plant and 25 metres from both.

(b) Show the new position of the treasure on your scale drawing.

Exercise 26.3 — Links: (*26D*) 26D

1 Triangle ABC is similar to triangle XYZ. AB = 5 cm, XY = 10 cm, AC = 8 cm. Work out the length of XZ.

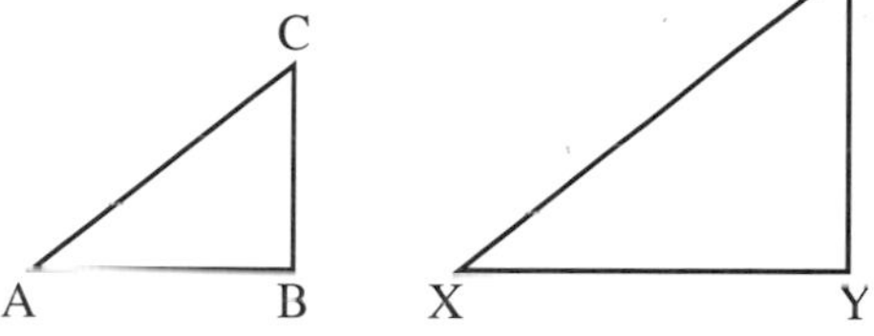

2 Two similar statues have heights of 10 metres and 15 metres respectively. The length of the arm on the larger statue is 3 metres. Work out the length of the arm on the smaller statue.

3 AB = 5 cm, BC = 2 cm, BE = 4 cm, DE = 3 cm.

(a) Calculate the length of CD.

(b) Calculate the length of AD.

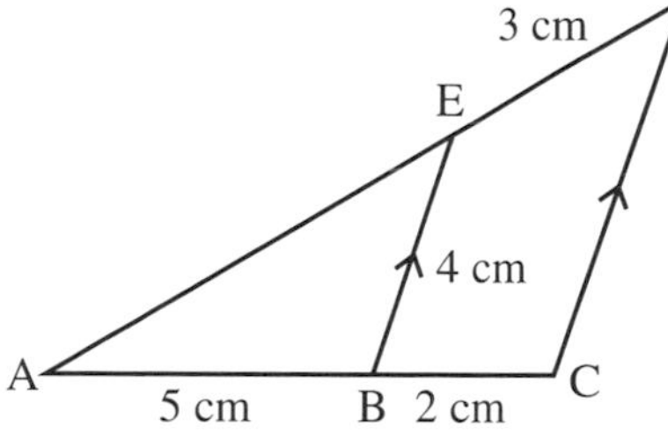

4 The height of a small ice-cream cone is 120 mm. It has a diameter of 40 mm. The diameter of the large ice-cream one is 50 mm. Calculate the height of the large ice-cream cone.

5 In each pair of similar diagrams, calculate the lengths of the sides marked with a letter:

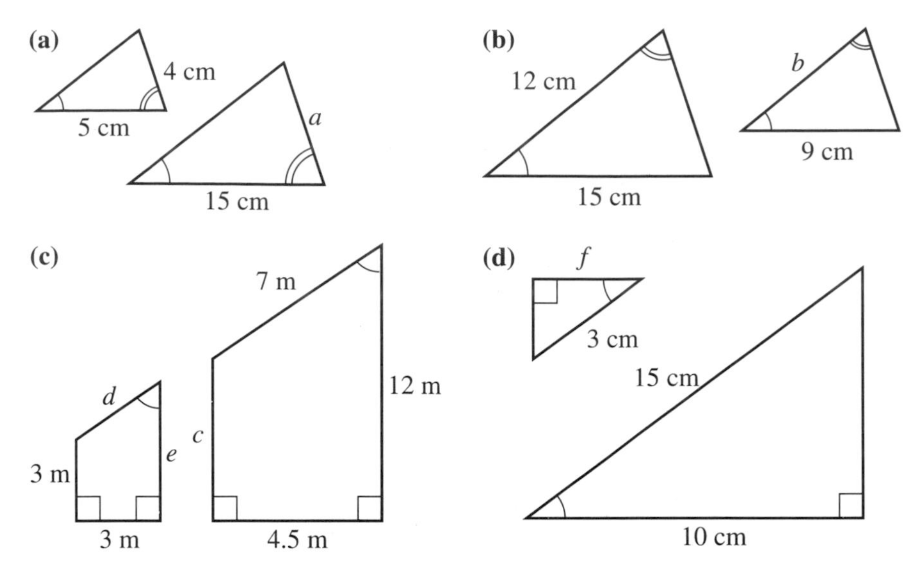

6 **(a)** Explain why these two triangles are similar.

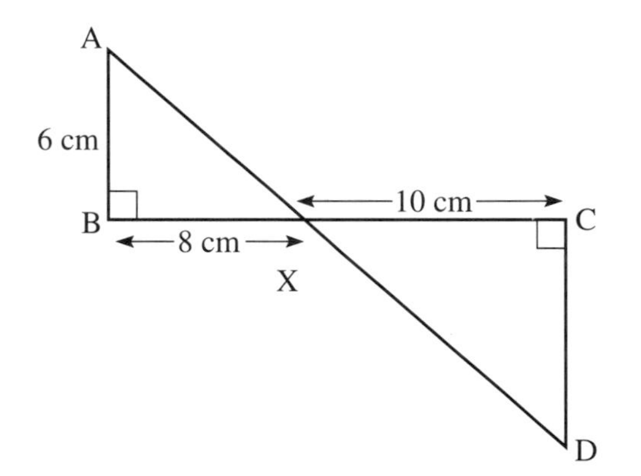

(b) Find all of the missing lengths in the triangles.

7 **(a)** Explain why these two shapes are similar.

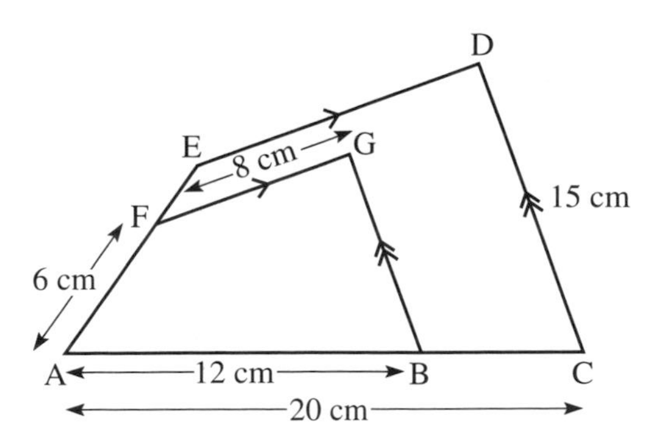

(b) Find all the missing lengths in each quadrilateral.

Exercise 26.4 Links: (*26E*) 26E

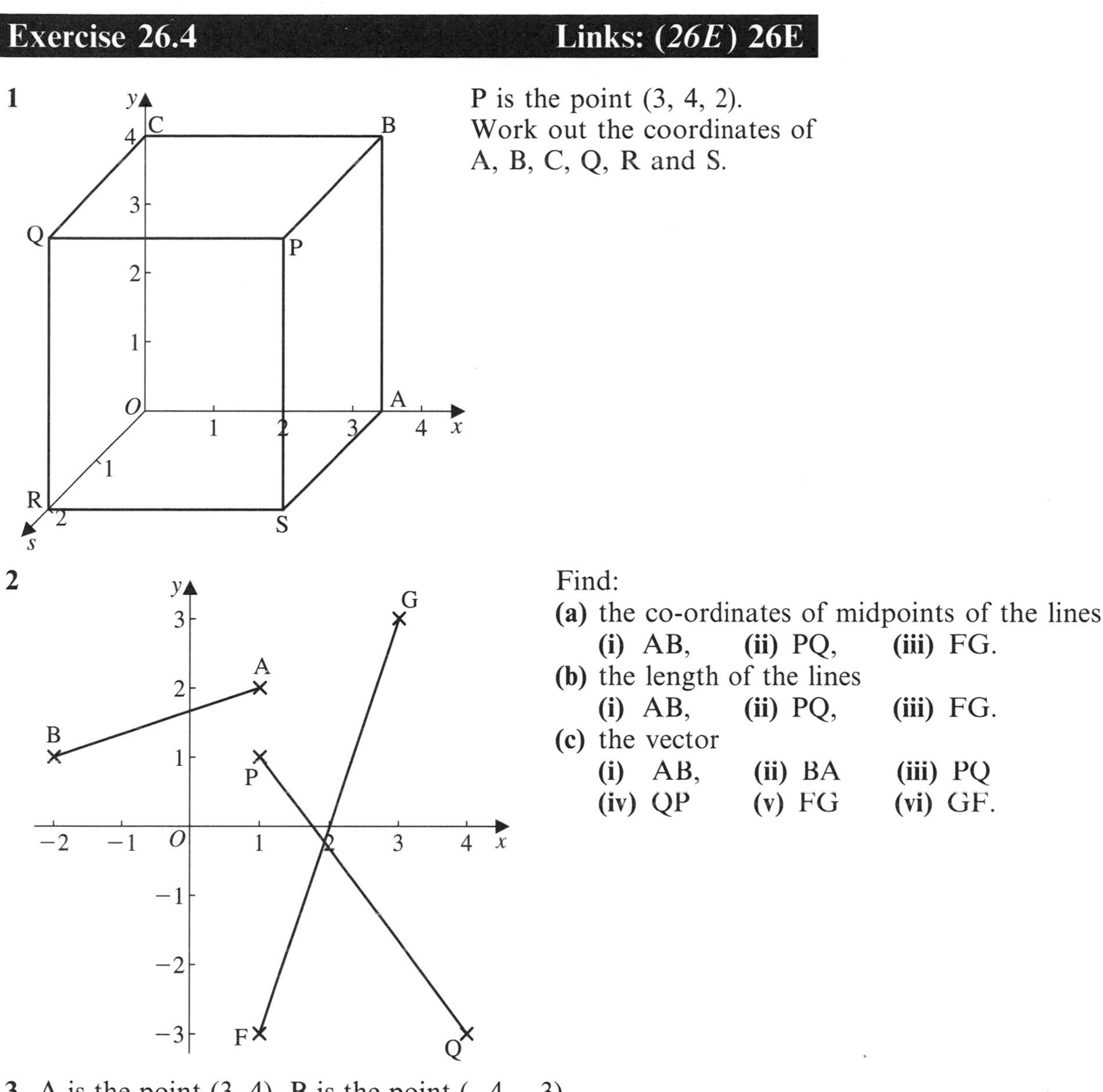

1 P is the point (3, 4, 2).
Work out the coordinates of A, B, C, Q, R and S.

2 Find:

(a) the co-ordinates of midpoints of the lines
  (i) AB, **(ii)** PQ, **(iii)** FG.

(b) the length of the lines
  (i) AB, **(ii)** PQ, **(iii)** FG.

(c) the vector
  (i) AB, **(ii)** BA **(iii)** PQ
  (iv) QP **(v)** FG **(vi)** GF.

3 A is the point (3, 4), B is the point (−4, −3).

(a) Find the coordinate of P, the mid point of AB.

(b) Work out the length of AB.

(c) Write down the vector
  (i) AB **(ii)** BA **(iii)** AP.

Exercise 26.5 Links: (*26E*) 26F

1 The diagram shows the end section of a garden shed.

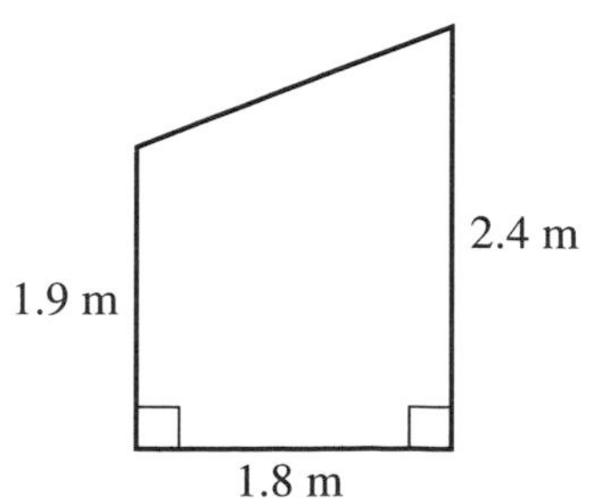

(a) Make a scale drawing of this end section. Use a scale of 5 cm to represent 1 m.

(b) Use your scale drawing to find
  (i) the angle the roof of the shed makes with the horizontal
  (ii) the actual length of the roof of the shed.

2 Mark two points P and Q, 5 cm apart. Shade in the locus of the points that are less than 3.5 cm from P **and** nearer Q than P.

3 PQ = 2 cm, PT = 1.6 cm, RS = 5 cm, RT = 4 cm.
(i) Work out the length of QT.
(ii) Work out the length of ST.

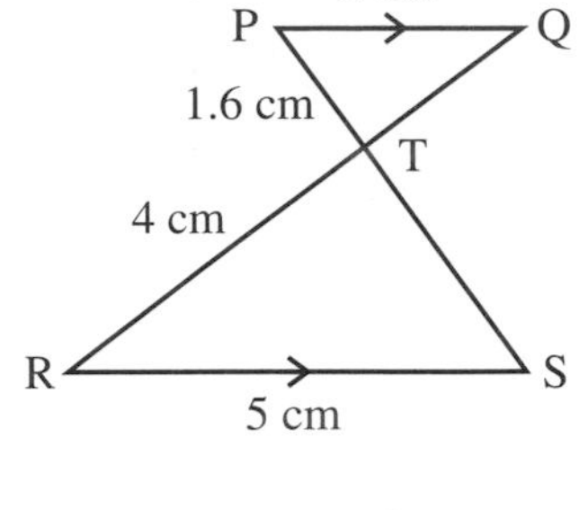

4 AD = 5 cm, BD = 2 cm, BC = 6 cm, AE = 4.5 cm.
(i) Work out the length of DE.
(ii) Work out the length of AC.

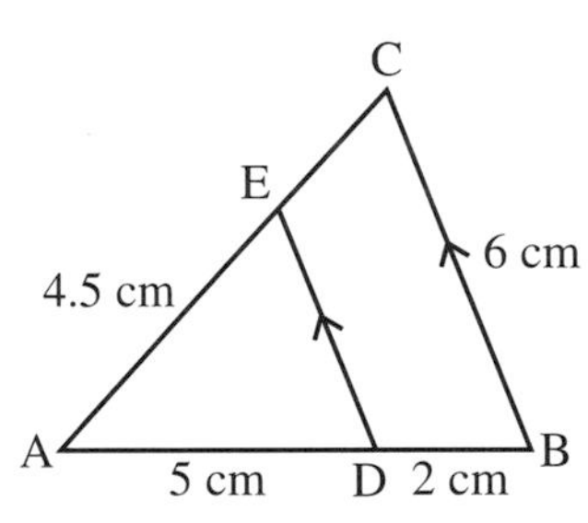

27 The sine and cosine ratios

Exercise 27.1 Links: (*27A, B*) 27A, B

You will need a ruler and protractor.

1 **(a)** Construct accurately 3 different sized right-angled triangles where angle CAB = 65°.

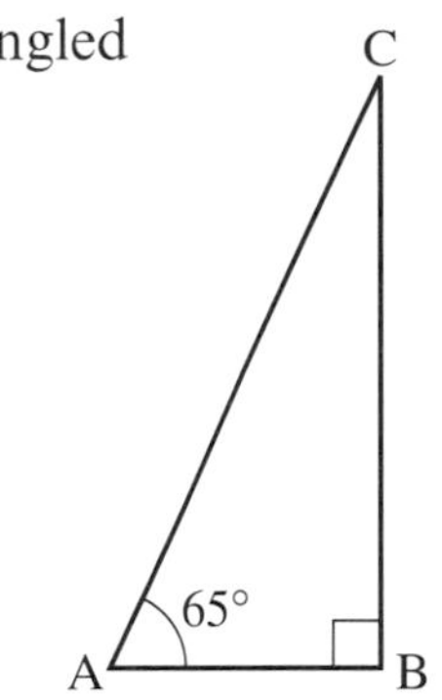

For each triangle:

(b) measure the lengths CB and AC

(c) work out the ratio $\frac{\text{CB}}{\text{AC}}$.

(d) What do you notice about the 3 ratios $\frac{\text{CB}}{\text{AC}}$?

2 Use your calculator to find the value of:

(a) sin 27°
(b) sin 55°
(c) sin 79°

3 Use your calculator to work out the value of θ:

(a) $\sin\theta = 0.3$
(b) $\sin\theta = 0.669$
(c) $\sin\theta = 0.9925$

4 Calculate the named lengths in these triangles:
Give your answers correct to 3 s.f.

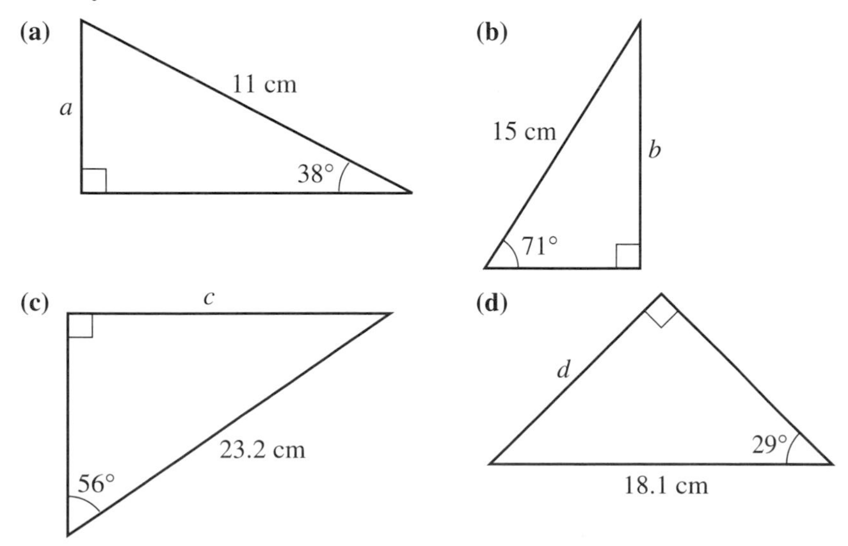

5 In these triangles find the named angles. Give your answers correct to 3 s.f.

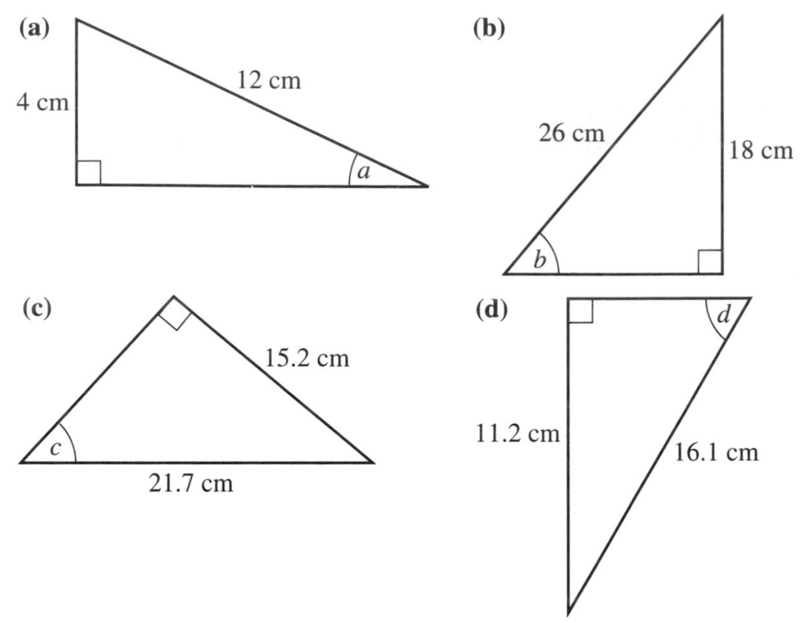

Exercise 27.2 Links: (27C–F) 27C–F

You will need a ruler and protractor.

1 **(a)** Construct accurately 3 different sized right-angled triangles where CAB = 35°.

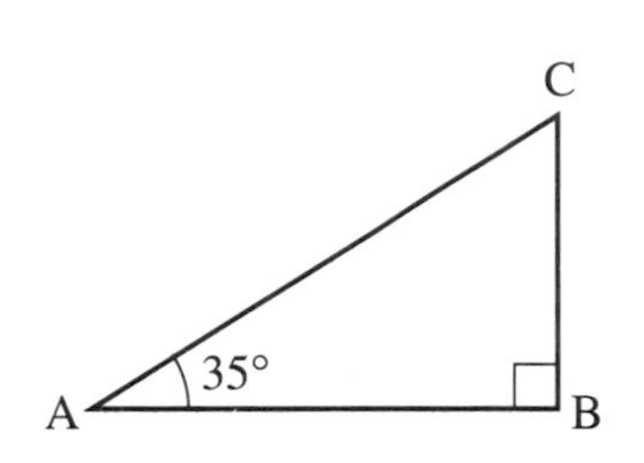

For each triangle:

(b) measure the length AB and AC

(c) work out the ratio $\frac{AB}{AC}$.

(d) What do you notice about the 3 ratios $\frac{AB}{AC}$?

2 In this question use a calculator and write your answers correct to 4 d.p. Copy and complete the table:

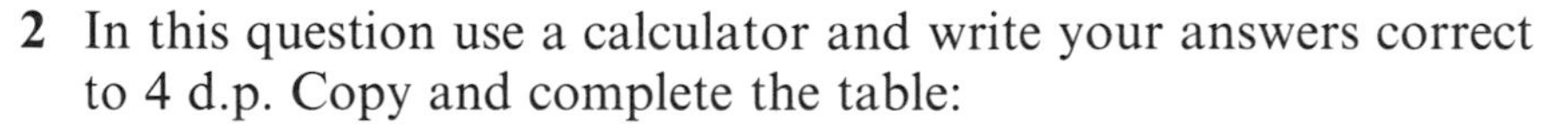

θ	$\sin\theta$	$(90° - \theta)$	$\cos(90° - \theta)$
15°	0.2588	75°	0.2588
28°			
35°			
42°			
55°			

In the following questions give your answers correct to 3 d.p.

3 Use your calculator to find the value of:

(a) cos 17° **(b)** cos 36° **(c)** cos 88°

4 Use your calculator to find the value of θ:
(a) $\cos\theta = 0.174$ **(b)** $\cos\theta = 0.4$ **(c)** $\cos\theta = 0.5592$

5 Calculate the named lengths in these triangles:

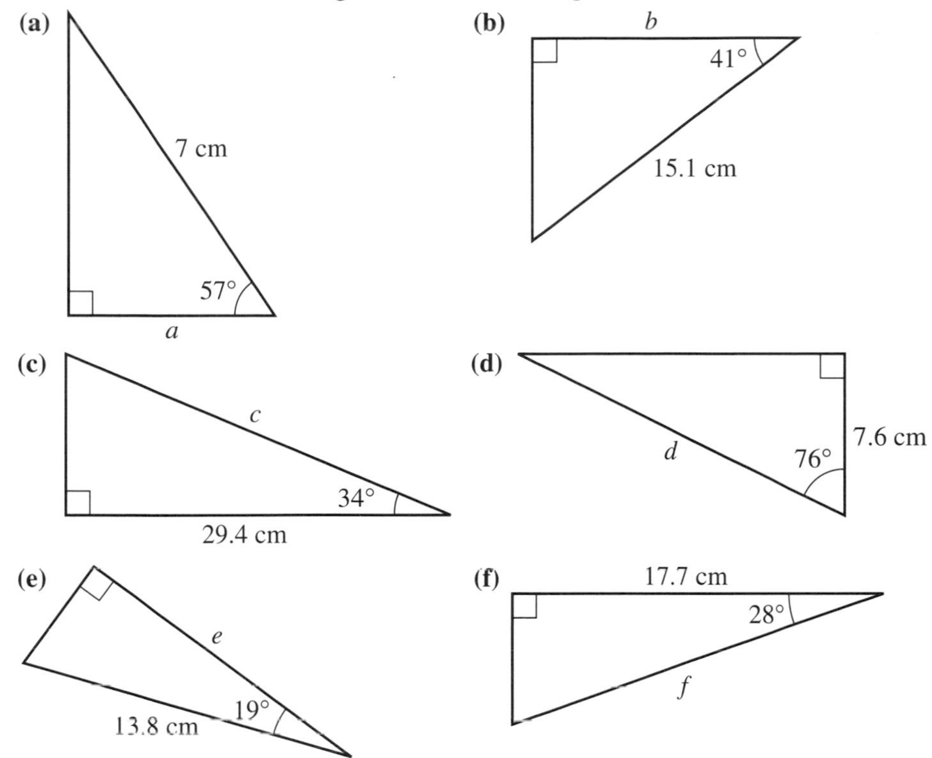

6 Calculate the named angles in these triangles:

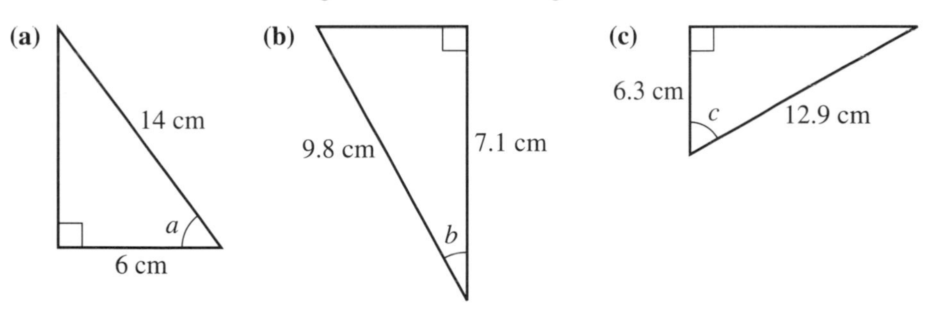

7 Sonia can see a buoy from the top of a cliff. The angle of depression is 38°. Sonia is 38 m above sea level.
Calculate the straight line distance from Sonia to the buoy.

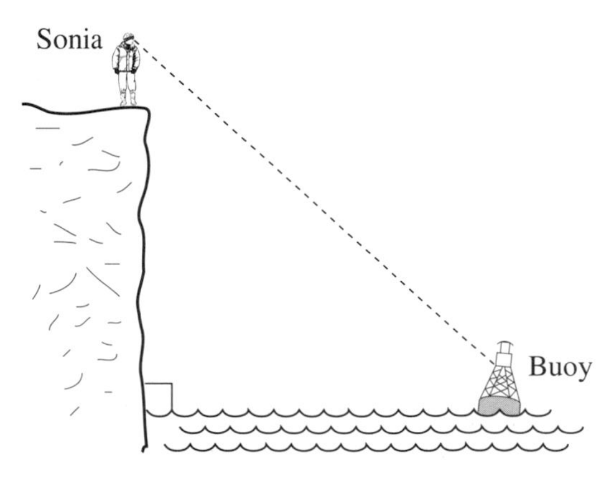

8 George, an electricity worker, climbs a pylon to repair cables. The pylon is 18 m high. His workmate Harry is holding the end of a cable between himself and George that is 25 m long. Calculate the angle of elevation from Harry to George.

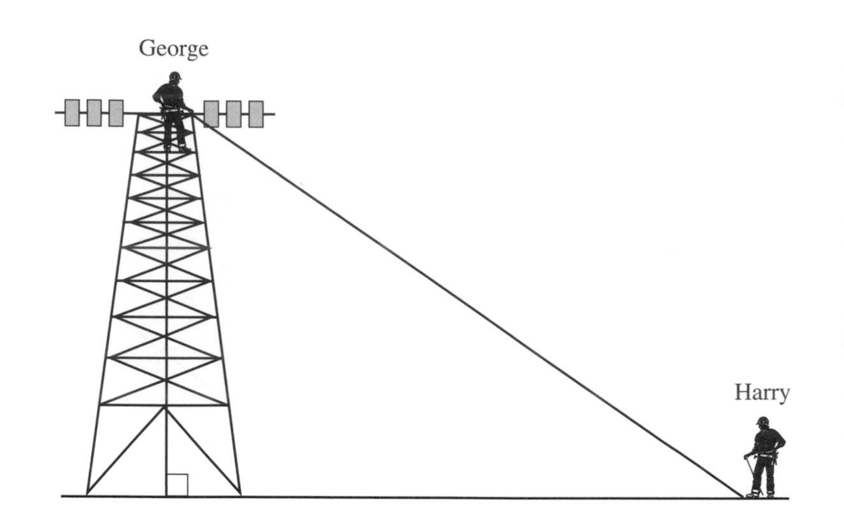

Exercise 27.3 Links: (*27G*) 27G

1 A ladder 11 m long is leaning against a vertical wall. It just reaches the gutter. The ladder makes an angle of 65° with the horizontal ground.
Calculate the height of the gutter above the ground.

2 In the diagram ACD and BCD are right-angled triangles.
Calculate:
(a) DC
(b) AC
(c) AB
(d) angle BDC
(e) angle ADB

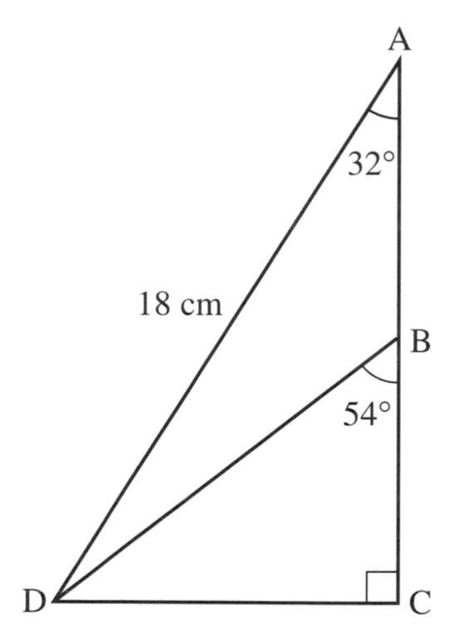

3 A cable used to support a pole makes an angle of 56° with the horizontal. The cable is 10.5 m long.
Calculate the height of the cable above the ground where it meets the pole.

4 Tina stands at the top of a tower. Emma, her friend, is on the ground 45 m from the bottom of the tower. The angle of depression of Emma from Tina is 43°.
Calculate the height of the tower.

5 A flagpole casts a shadow 21.5 m long. The altitude of the sun is 52°. What is the height of the flagpole?

28 Equations and inequalities

Exercise 28.1 **Links: (*28A, B*) 28A, B**

1 A martial arts club charge customers:

£50 to join the club and then
£3 for every session attended

Mushtaq joins the club and attends x sessions. The total charge to Mushtaq for this is £C.

(a) Write down a formula for C in terms of x.
(b) Draw the number machine for this formula.
(c) Draw the inverse number machine.
(d) Use the inverse number machine to work out the value of x in the cases when the total charge, £C, is:
(i) £80 **(ii)** £350 **(iii)** £146
(iv) £191 **(v)** £236 **(vi)** £293
(e) Explain why the total cost could not be exactly £100.

2 Fiona works as a cosmetics consultant. She travels around the country. She is paid on a daily basis. Her company pay her:

£80 per day plus
40 p for each mile she travels per day

(a) Write this as a formula using P for her daily pay in pounds and x for the number of miles she travels per day.
(b) Draw a number machine for your formula.
(c) Draw the inverse number machine.
(d) Use the inverse number machine to work out how many miles Fiona travelled each day when her pay for the day was:
(i) £120 **(ii)** £92 **(iii)** £107.20
(iv) £140 **(v)** £142.80 **(vi)** £180.40

3 Solve these equations:

(a) $47 = 5m + 2$ **(b)** $48 = 4n + 20$ **(c)** $37 = 5p - 3$

(d) $40 = 4(x + 1)$ **(e)** $8 = 2(y - 3)$ **(f)** $3 = \frac{b}{4} + 8$

(g) $3h - 5 = 7$ **(h)** $5x + 2 = 22$ **(i)** $6x - 1 = 11$

(j) $7 + x = 9$ **(k)** $\frac{a + 3}{4} = 2$ **(l)** $\frac{2b - 3}{5} = 7$

(m) $\frac{m}{4} - 1 = 4$ **(n)** $\frac{3n - 6}{4} = 1\frac{1}{2}$ **(o)** $\frac{4a + 2}{6} = 7$

(p) $\frac{x}{4} + 1 = 1\frac{1}{4}$ **(q)** $3y - 1 = 1$ **(r)** $\frac{1}{2}n - 2 = 3$

(s) $3a + 7 = 7$ **(t)** $\frac{4y + 1}{5} = 3$ **(u)** $\frac{2x}{3} - 1 = 2$

(v) $5y - 2 = 18$ **(w)** $\frac{3 + 5a}{11} = 3$ **(x)** $2 = \frac{6b}{7} - 1$

Exercise 28.2 Links: (*28C, D*) 28C, D

1 Solve each of these equations:

(a) $3x + 1 = 7$ **(b)** $4y - 2 = 7$ **(c)** $7n - 1 = 12$

(d) $\frac{a}{5} + 2 = 3$ **(e)** $\frac{3b}{8} - 5 = 2$ **(f)** $6(x - 1) = 18$

(g) $4(3x + 2) = 32$ **(h)** $\frac{m + 5}{6} = 4$ **(i)** $5\left(\frac{x}{4} - 2\right) = 20$

(j) $\frac{1}{2}(3x - 5) = 8$ **(k)** $\frac{1}{4}(4y + 2) = 4\frac{1}{2}$ **(l)** $\frac{3n - 2}{5} = 2$

2 Solve each of these equations:

(a) $4x + 3 = 3x + 7$ **(b)** $2y - 3 = y + 2$

(c) $5n + 2 = 3n + 6$ **(d)** $4m + 7 = m + 19$

(e) $3 + 2x = x + 8$ **(f)** $5y + 8 = 7y - 6$

(g) $11x + 3 = 5x + 21$ **(h)** $3n - 12 = 8n + 3$

(i) $4y + 1 = 2y - 5$ **(j)** $5x - 1 = 2x - 7$

3 Mrs Cundy needs to have her central heating system serviced.
She can choose from two local companies.

Ficsit charge: *A call out charge of £25 and then £8 per hour working on the job.*

Mendems charge: *A call out charge of £37 and then £6 per hour working on the job.*

Each company takes the same amount of time to do the job.
Using x to represent the number of hours the job takes:

(a) write down expressions for the total charges made by the two companies

(b) write down and solve an equation which gives the time taken for the two total charges to be equal.

4 The triangle and pentagon each have the same perimeter. Write down an equation and solve it to find the value of x.

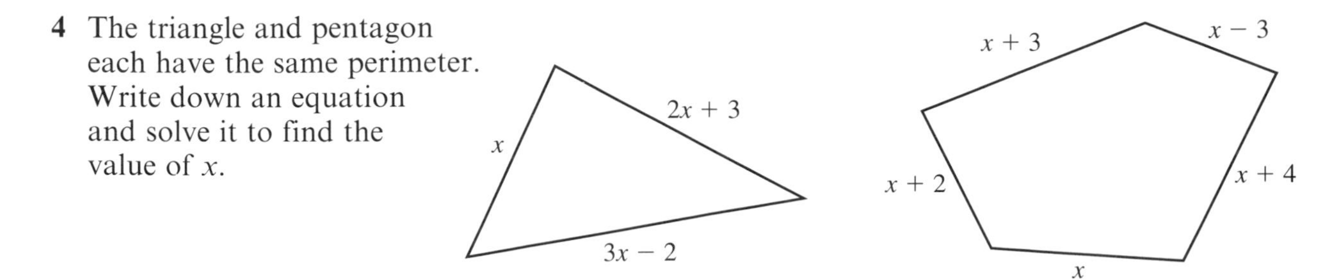

5 Solve the equations:

(a) $6x - 2 = 3x + 13$ **(b)** $x + 3 = \frac{1}{2}x - 4$

6 Alfa Autos and Betta Cars are two car hire firms. Alfa Autos charge £50 per day and 20 p per mile. Betta Cars charge £30 per day and 40 p per mile. Karen wishes to hire a car for a day from one of these companies.
How far will she travel if the two charges are equal?

Exercise 28.3 Links: (*28E, F*) 28E, F

1 Solve each of these equations:

(a) $3x + 7 = 22 - 2x$ **(b)** $6x - 3 = 13 - 2x$

(c) $5y - 4 = 26 - 5y$ **(d)** $1\frac{1}{2}x + 2 = 8 - \frac{1}{2}x$

(e) $4a + 1 = 8 - 3a$ **(f)** $8n + 11 = 1 + 6n$

(g) $3 - 4x = 12 - 7x$ **(h)** $y + 1 = 4 - 2y$

(i) $17 + m = 31 - 6m$ **(j)** $6 - x = 11 - 3x$

(k) $5 - 4x = 3 + x$ **(l)** $7 + 3y = 5y - 2$

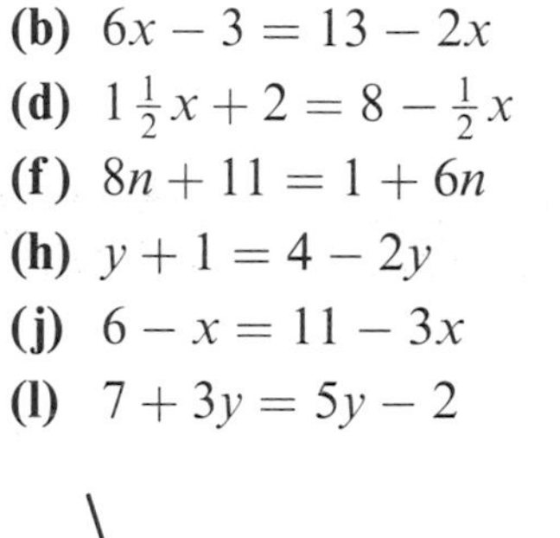

2 The diagram shows two of the sides of a regular pentagon.

(a) Write down an equation in x which must be true.

(b) Solve this equation.

(c) Work out the perimeter of the pentagon.

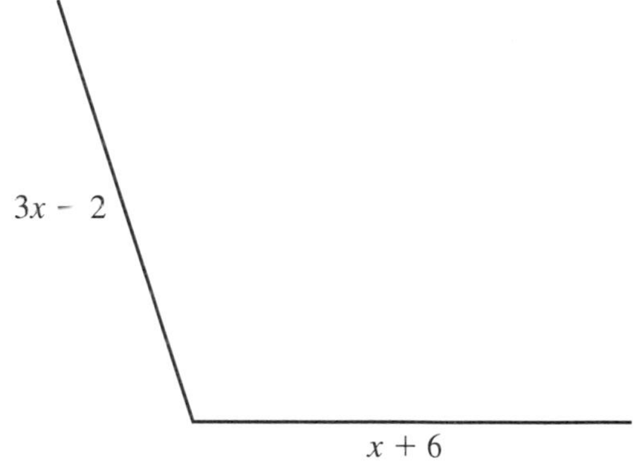

3 At the start of the week, Alex has £20 and Mary has £17. During the week Alex hires four videos and Mary hires two videos. Each video costs the same amount to hire. Neither of them spends any other money. At the end of the week they have the same amount of money left. Let x stand for the cost, in £, of hiring a video.

(a) Write down an equation in x which must be true.

(b) Solve the equation to find the cost of hiring a video.

4 Solve each of these equations:

(a) $3(x + 1) = 15$ **(b)** $2(y + 3) = y + 11$

(c) $4(n - 3) = 3(n + 1)$ **(d)** $5(x + 2) = 3x + 16$

(e) $7x - 3 = 3(2x + 1)$ **(f)** $4y = \frac{1}{2}(2y + 3)$

(g) $5 + 3x = 2(2x + 3)$ **(h)** $3(1 - 2y) = 8$

(i) $5(3n + 1) = 12n + 14$ **(j)** $3(4 - 5y) = 4(6 - 5y)$

(k) $7(9 - y) = 5(3y - 5)$ **(l)** $4(5 + x) = 5(x - 3) + 22$

5 Solve each of these equations:

(a) $\frac{x}{2} + 3 = \frac{x}{4} + 5$ **(b)** $\frac{y}{3} + 3 = \frac{y}{4} + 4$

(c) $2x - 1 = \frac{3x}{2} + 6$ **(d)** $\frac{n+5}{3} = \frac{n}{7} + 3$

(e) $\frac{x}{3} = 5(x - 2)$ **(f)** $\frac{3a}{4} - 1 = 5(a - 1)$

(g) $\frac{b+1}{5} = \frac{b-1}{4}$ **(h)** $5(3x - 2) = \frac{2x+3}{2}$

6 Solve each of these equations:

(a) $7 - 4x = 15$ **(b)** $3(2y - 1) = y$ **(c)** $4(n - 3) = \frac{n}{5}$

Exercise 28.4 Links: (*28G–J*) 28G–J

1 The diagram shows the graphs of $y = 3x + 2$ and $y = 12 - x$.

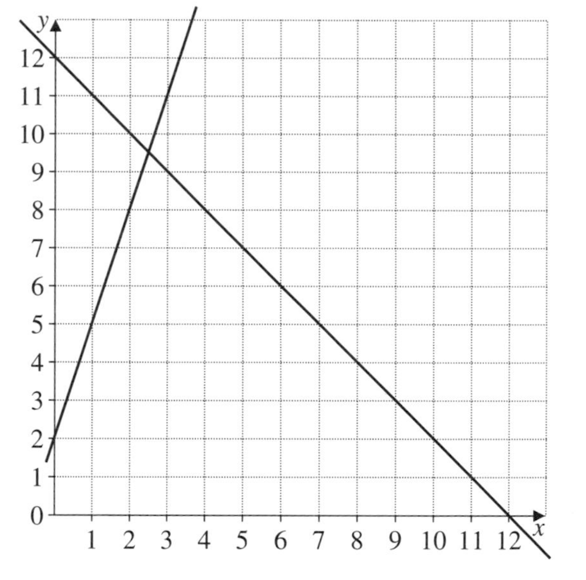

Use the graphs to find the solutions of the simultaneous equations $y = 3x + 2$ and $y = 12 - x$.

2 **(a)** Draw the graphs of $y = 3x + 1$ and $y = 7 - x$ for values of x from 0 to 5.

(b) Use your graphs to find the solution to the simultaneous equations $y = 3x + 1$ and $y = 7 - x$.

3 **(a)** Draw the graphs of $y - x = 6$ and $y = -x + 1$ for values of x from -6 to 2.

(b) Use your graphs to solve the simultaneous equations $y - x = 6$ and $y = -x + 1$.

4 Janet and Liz are two sisters.
At the start of the year Janet has £360 and spends £5 per week.
At the start of the same year Liz has no money but saves £5 per week. Neither sister either gets or spends any other money.

(a) Write down expressions, in terms of n, for the amount of money each sister has n weeks after the start of the year.

(b) Draw graphs of the amount of money each sister has for values of n from 0 to 52 weeks.

(c) Use your graphs to find the value of n when the two sisters have equal amounts of money.

5 Solve by substitution these pairs of simultaneous equations:

(a) $y = 25 + 2x$, $y = 5x - 2$ **(b)** $y = 3x + 20$, $y = 7x$ **(c)** $y = 4x - 7$, $y = 2x + 1$

(d) $y = 3x - 5$, $y = 8x + 5$ **(e)** $y = 5 - 2x$, $y = x + 17$ **(f)** $y = 7 - 2x$, $y = 15 - 3x$

(g) $y + 5 = 3x$, $y = x + 1$ **(h)** $2y = x + 1$, $y = x - 3$ **(i)** $y - 5 = 3x$, $2y = 7x + 9$

(j) $y + 2x = 11$, $3x - y = 9$ **(k)** $y - 3x = 6$, $x + y = 2$ **(l)** $x + 3y = -1$, $2y + 3x = 11$

6 Solve these simultaneous equations by elimination:

(a) $x + y = 7$, $x - y = 1$ **(b)** $2x + y = 9$, $x - y = 6$ **(c)** $3x - y = 7$, $2x + y = 3$

(d) $7x - 2y = 25$, $3x + 2y = 25$ **(e)** $4x + 3y = 1$, $x - 3y = -11$ **(f)** $2n - 5m = 3$, $5m + n = 9$

(g) $5x + y = 23$, $2x + y = 11$ **(h)** $3p + 2q = 7$, $5p + 2q = 9$ **(i)** $2x + 3y = 3$, $5x + 3y = 12$

(j) $4a - b = 11$, $3a - b = 8$ **(k)** $3m - 4n = -18$, $5m - 4n = -22$ **(l)** $3y - 2x = 5$, $5y - 2x = 7$

(m) $3x + 2y = 7$, $4x - y = 13$ **(n)** $4x + 3y = 23$, $x + 5y = 27$ **(o)** $4a - 3b = -6$, $a + 2b = 1$

(p) $3x + 5y = 13$, $7x - 2y = 3$ **(q)** $4m - 3n = 11$, $5m + 2n = 8$ **(r)** $7a - 3b = 10$, $3a - 7b = 10$

(s) $4x + 3y = 11$, $3x + 2y = 8$ **(t)** $5a + 3b = 21$, $4a + 5b = 22$ **(u)** $3x - 4y = 10$, $2x + 7y = -5$

(v) $x + 2y = 7$, $2y - x = 5$ **(w)** $3n - 2m = -10$, $m + 2n = 2$ **(x)** $4a + 3b = 10$, $4b + 3a = 11$

7 A group of 4 adults and 3 children go to the cinema.
The total cost of their tickets is £35.
A second group of 6 adults and 5 children go to the same cinema.
The total cost of their tickets is £54. The cost of an adult's ticket was x pounds. The cost of a child's ticket was y pounds.

(a) Use the information to write down two simultaneous equations in x and y.

(b) Solve these equations to find the cost of:
(i) an adult's ticket **(ii)** a child's ticket

8 Mary thought of two numbers.
She doubled her first number, trebled her second number, added these two results together and got 5.
Then she multiplied her first number by 4 and added this to her second number. This time she got a result of 15.
(a) By writing x for Mary's first number and y for her second number, write down two simultaneous equations in x and y.
(b) Solve these equations to work out the two numbers which Mary thought of originally.

9 Solve the simultaneous equations:

$$3x - 5y = 21$$
$$5x + 3y = 1$$

Exercise 28.5 Links: (*28K*) 28K, L, M

1 Write down all the integers which satisfy each inequality:
(a) $2 < x < 9$ **(b)** $-3 < x < 5$ **(c)** $5 < n \leqslant 12$
(d) $-4 < m < 0$ **(e)** $-1 \leqslant n < 2$ **(f)** $-3 \leqslant n \leqslant 3$
(g) $-6 < 2x < 10$ **(h)** $-9 < 3n \leqslant 12$ **(i)** $-5 \leqslant 2m \leqslant 12$

2 Solve these inequalities:
(a) $4n < 12$ **(b)** $3n \leqslant 18$ **(c)** $5n < 10$
(d) $5n > 20$ **(e)** $4n > -12$ **(f)** $7n \geqslant 21$
(g) $2n \leqslant -8$ **(h)** $4n < -2$ **(i)** $6n \leqslant -2$
(j) $5n \geqslant -15$ **(k)** $2n \geqslant 9$ **(l)** $4n \geqslant -3$

3 Solve these inequalities:
(a) $5n + 3 > 23$ **(b)** $x - 7 < 1$ **(c)** $2x - 3 \geqslant 4$
(d) $3n + 7 \leqslant -2$ **(e)** $1 + 3n > 13$ **(f)** $5x + 7 < 32$
(g) $3x + 1 < 2x$ **(h)** $3x - 2 > 2x + 5$ **(i)** $4n + 3 \leqslant 2n + 7$
(j) $5x - 2 \geqslant 2x + 10$ **(k)** $3 - 2x < 4x - 7$ **(l)** $3x + 9 < 5x - 1$

4 Solve each of these inequalities. Show each solution as a shaded region on a graph.
(a) $5x - 3 \leqslant 17$ **(b)** $4y + 1 > 9$
(c) $3x - 2 \leqslant x + 8$ **(d)** $3y + 4 > 2y + 1$

5 Shade a region to represent each inequality:
(a) $y + 2x \leqslant 1$ **(b)** $3y + 2 > x$
(c) $3x + 4y < 12$ **(d)** $2y - x \geqslant 7$

6 Shade a region which satisfies *both* inequalities:
(a) $y < 2x + 1$, $x < 0$ **(b)** $x + y > 5$, $x - y < 7$
(c) $3x + 4y < 12$, $y > 0$ **(d)** $y > 7 - 2x$, $x + 3y > 12$

7 A courier company has £240,000 to buy new vehicles.
They will buy cars and motorcycles.
Each car costs £12,000.
Each motorcycle costs £4000.
The company has space for a maximum of 50 new vehicles.
What are the possible combinations of vehicles they could buy?
Show your answer as a solution set on a graph.

Exercise 28.6 Links: (*28L*) 28N

1 Solve each of these quadratic equations:

(a) $x^2 - 8x + 15 = 0$ **(b)** $x^2 - 10x + 24 = 0$
(c) $x^2 - 7x + 10 = 0$ **(d)** $x^2 + 7x + 12 = 0$
(e) $x^2 + 8x + 15 = 0$ **(f)** $x^2 - 16x - 17 = 0$
(g) $x^2 - 2x - 24 = 0$ **(h)** $x^2 - 3x - 15 = 0$
(i) $x^2 - 7x = 0$ **(j)** $x^2 + 5x = 0$
(k) $x^2 - 36 = 0$ **(l)** $100 - x^2 = 0$
(m) $x^2 - 5x = -6$ **(n)** $x^2 + 7x = -12$
(o) $x^2 - 2x - 35 = 0$ **(p)** $x^2 + 2x - 35 = 0$
(q) $2x^2 = 18$ **(r)** $72 = 3x^2$
(s) $x^2 - 9x + 14 = 0$ **(t)** $x^2 - 4x - 21 = 0$
(u) $n(n + 2) - 35$ **(v)** $n(n - 2) = 120$
(w) $n(n - 3) = 40$ **(x)** $n^2 + n = 20$
(y) $p^2 - 8p + 16 = 0$ **(z)** $q^2 + 10q + 25 = 0$

2 The diagram shows a rectangle ABCD.
AB = x cm, BC = $(x - 2)$ cm.
The area of ABCD = 48 cm^2.
(a) Show that $x^2 - 2x - 48 = 0$.
(b) Solve this equation to find the value of x.

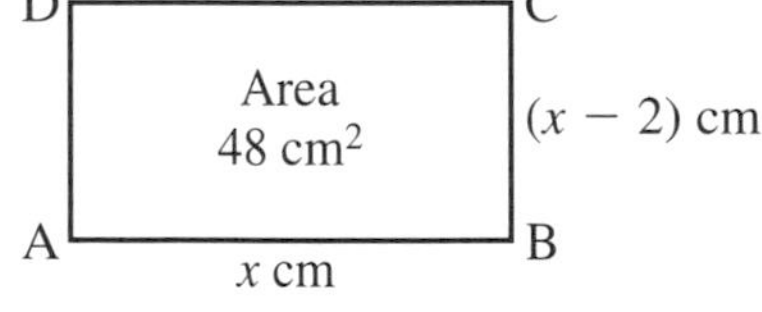

3 PQR is a triangle of area 20 cm^2.
The angle at Q = 90°.
PQ = x cm and QR is 6 cm longer than PQ.
(a) Show that $x^2 + 6x - 40 = 0$.
(b) Solve this equation to find the value of x.

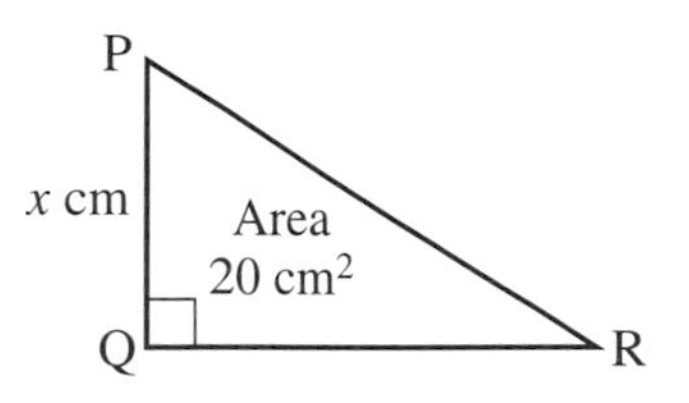

Exercise 28.7 Links: (*28M*) 28O

1 Solve the equations:

(a) $3y + 7 = 28$ **(b)** $2(3p + 2) = 19$ **(c)** $3t - 4 = 5t - 10$

[E]

2 **(a)** Draw the graphs of
(i) $x + y = 4$ **(ii)** $y = x + 2$
for values of x from -3 to 3.
(b) Use the graphs to solve the simultaneous equations

$$x + y = 4$$
$$y = x + 2$$

[E]

3 x is an integer, such that $-3 < x \leqslant 2$
List all the possible values of x. [E]

4 George repairs electrical goods. He charges:

the cost of the spare parts plus
£8 for every hour that he works on a job

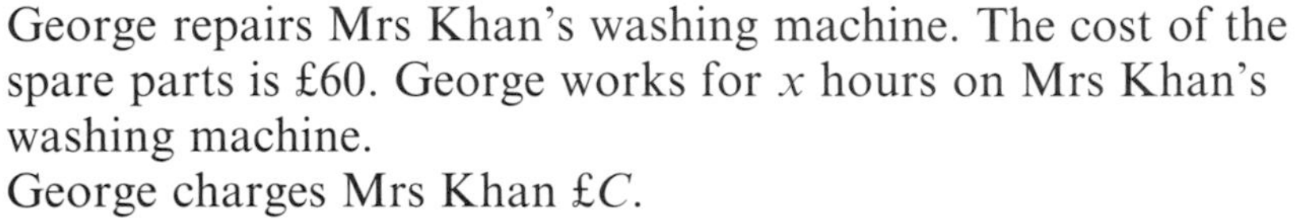
George repairs Mrs Khan's washing machine. The cost of the spare parts is £60. George works for x hours on Mrs Khan's washing machine.
George charges Mrs Khan £C.
(a) Write down a formula connecting x and C.

George charges Mrs Khan £92.
(b) Work out how many hours George works on Mrs Khan's washing machine. [E]

5 Solve the simultaneous equations:
(a) $3x - y = 7$
$2x + y = 3$
(b) $4p + 3q = 9$
$2p - q = 7$
(c) $3a + 2b = -1$
$2a + 5b = 14$

6 Solve the quadratic equation

$$x^2 + 2x - 24 = 0$$

[E]

7 Solve each of the inequalities:
(a) $4x - 7 < 25$ **(b)** $2x + 1 \geqslant 17$ **(c)** $4 - 3x > 9 - 5x$

8 PQRS is a rectangle.
PQ = x cm.
PS is 4 cm shorter than PQ.
(a) Write down an expression, in terms of x, for the length of PS.

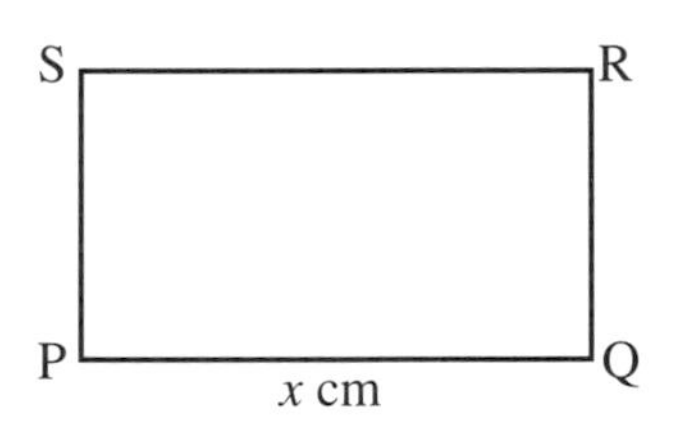

The area of PQRS is 45 cm.
(b) Show that $x^2 - 4x - 45 = 0$.
(c) Solve this equation to find the value of x.

9 **(a)** On a graph, shade the region defined by the inequality

$$0 < x < 4$$

(b) On the same axes, shade the region for which

$$y > 1$$

(c) On the same axes again, draw the line

$$x + 3y = 12$$

(d) Shade the region for which

$$x + 3y < 12, \quad 0 < x < 4 \quad \text{and} \quad y > 1$$

10 **(a)** Solve each of the equations

(i) $3(2x - 1) = 21$ **(ii)** $7 - 3x = 11$

(b) Solve the simultaneous equations

$2x - 3y = 9 \qquad 3x + y = 8$

(c) Solve the equation

$x^2 - 3x - 40 = 0$

11 The number of diagonals inside a regular polygon with n sides is given by

$$\text{number of diagonals} = \frac{n(n-3)}{2}$$

(a) Show that if a regular polygon with n sides has 65 diagonals then

$n^2 - 3n - 130 = 0$

(b) Solve the above equation to find the value of n.

29 Interpreting data

Exercise 29.1 Links: (*29A*) 29A

1 Two holiday resorts claim to be the sunshine capital. Resort A claims a daily average of 10.7 hours for May–August with a maximum of 14.3 hours. Resort B claims a daily average of 11.1 hours for June–July with a maximum of 14.2 hours. Which resort do you think should be top and why?

2 Two cars are tested on fuel consumption. Each car makes the same 10 journeys of 25 miles. The results are in the table:

	Fuel consumption in litres.									
Car F	4.1	4.7	4.2	4.3	4.1	4.5	4.2	4.1	4.1	4.3
Car G	4.2	4.4	4.4	4.2	4.2	4.3	4.2	4.3	4.2	4.2

Make appropriate statistical comparisons to decide whether or not there is evidence that one is better than the other.

3 Two athletes, Quick and Fleet, are competing in a 200 m race. Their times for their last six races are in the table.

Quick	21.2	21.3	21.2	21.3	21.2	21.3
Fleet	21.7	20.9	21.3	21.3	21.4	21.5

The commentator says that Fleet had the fastest time this season and is expected to win. Say, with reasons, whether you agree or not.

4 The two tables are of traffic surveys taken in North Street and South Street. Give two reasons why you cannot be sure that North Street is busier than South Street.

North Street	
Cars	73
Buses	2
Vans	11
Other	8
Total	94

South Street	
Cars	21
Buses	1
Vans	3
Other	4
Total	29

5 There are 50 pupils in each of the groups, Year 9, Year 10, Year 11 at Lucea High School. A survey was carried out to find how many pets these pupils owned. The results are shown in the table.

Number of pets	Year 9	Year 10	Year 11
0	1	5	32
1	29	22	11
2	14	19	6
3	5	4	1
4	1	0	0

(a) What is the most common number of pets owned?
(b) Which year group had the least number of pets? Use the figures to give reasons for your answer. [E]

6 Rabbits were first taken to Australia in the middle of the 19th century. The estimated number of rabbits in Australia during the years 1850 to 1990 is recorded in the table below.

Year	1850	1870	1890	1910	1930	1950	1970	1990
Millions of rabbits	1.1	2.3	4.9	7.8	13.2	17.4	8.7	9.4

Describe briefly how the number of rabbits changed during these years. [E]

7 A survey was carried out in Mathstown High School to find out how long it takes the students to travel to school. The results of the survey are shown in the table.

Time, t minutes to travel to school	Number of Year 9 students	Number of Year 10 students	Number of Year 11 students
$0 < t \leqslant 10$	23	15	14
$10 < t \leqslant 20$	16	14	12
$20 < t \leqslant 30$	9	17	19
$30 < t \leqslant 40$	4	1	5
Totals	52	47	50

(a) **(i)** Use the information in the table to write down the year group which takes the longest time to travel to school.
(ii) Give a reason for your answer. [E]

8 In a factory there are 79 male and 74 female managers. Managers can either be junior or senior. There are 28 male senior managers. There is a total of 93 junior managers.
(a) Construct a two-way table to show the number of female and male managers in junior and senior management.
(b) Comment on the proportion of women in junior and senior management. [E]

Heinemann Educational Publishers
Halley Court, Jordon Hill, Oxford, OX2 8EJ
Part of Harcourt Education

Heinemann is the registered trademark of
Harcourt Education Limited

First published 2002

ISBN 0 435 53265 0

06 05 04 03
10 9 8 7

Designed and typeset by Techset, Tyne and Wear

Cover design by Miller, Craig and Cocking

Printed and bound by The Bath Press, Bath

Acknowledgements

The publisher's and author's thanks are due to Edexcel for permission to reproduce questions from past examination papers. These are marked with an [E]. The answers have been provided by the authors and are not the responsibility of Edexcel.